NATURAL RESOURCE MANAGEMENT

Volume 17

Trees, People and Power

Full list of titles in the set
Natural Resource Management

Volume 1: Genes from the Wild
Volume 2: Waterlogged Wealth
Volume 3: The Threatening Desert
Volume 4: Our Common Seas
Volume 5: Asking the Earth
Volume 6: The Rural Economy and the British Countryside
Volume 7: The Water Crisis
Volume 8: After the Green Revolution
Volume 9: Saving the Seed
Volume 10: Unwelcome Harvest
Volume 11: The Politics of Industrial Agriculture
Volume 12: Forest Politics
Volume 13: No Timber Without Trees
Volume 14: Controlling Tropical Deforestation
Volume 15: Plantation Politics
Volume 16: Saving the Tropical Forests
Volume 17: Trees, People and Power
Volume 18: Tropical Deforestation

Trees, People and Power

Peter Utting

earthscan
publishing for a sustainable future

London • Sterling, VA

First published in 1993

This edition first published in 2009 by Earthscan

ISBN 978-1-84971-028-2 (Volume 17)
ISBN 978-1-84971-011-4 (Natural Resource Management set)
ISBN 978-1-84407-930-8 (Earthscan Library Collection)

For a full list of publications please contact:

Earthscan
Dunstan House
14a St Cross Street
London EC1N 8XA, UK
Tel: +44 (0)20 7841 1930
Fax: +44 (0)20 7242 1474
Email: earthinfo@earthscan.co.uk
Web: **www.earthscan.co.uk**

22883 Quicksilver Drive, Sterling, VA 20166-2012, USA

Earthscan publishes in association with the International Institute for Environment and Development

A catalogue record for this book is available from the British Library

Library of Congress Cataloging-in-Publication Data has been applied for

Publisher's note
The publisher has made every effort to ensure the quality of this reprint, but points out that some imperfections in the original copies may be apparent.

At Earthscan we strive to minimize our environmental impacts and carbon footprint through reducing waste, recycling and offsetting our CO_2 emissions, including those created through publication of this book. For more details of our environmental policy, see www.earthscan.co.uk.

This book was printed in the UK by CPI Antony Rowe.
The paper used is FSC certified.

FSC
Mixed Sources
Product group from well-managed forests and other controlled sources
Cert no. SGS-COC-2953
www.fsc.org
© 1996 Forest Stewardship Council

Trees, people and power

The United Nations Research Institute for Social Development (UNRISD) is an autonomous agency that engages in multi-disciplinary research on the social dimensions of contemporary problems affecting development. Its work is guided by the conviction that, for effective development policies to be formulated, an understanding of the social and political context is crucial. The Institute attempts to provide governments, development agencies, grassroots organizations and scholars with a better understanding of how development policies and processes of economic, social and environmental change affect different social groups. Working through an extensive network of national research centres, UNRISD aims to promote original research and strengthen research capacity in developing countries.

Current research themes focus on the social dimensions of economic restructuring, environmental deterioration and conservation, ethnic conflict, the illicit narcotic drugs trade and drug control policies, political violence, the mass voluntary return of refugees, and the reconstruction of wartorn societies, as well as ways of integrating gender issues into development planning.

Peter Utting is a research coordinator at UNRISD, Geneva, specializing in social and political aspects of environmental programmes in developing countries and problems of reconstruction of wartorn societies. Based in Nicaragua throughout most of the 1980s, he has undertaken extensive research on food security and environmental issues in Central America. He is the author of *Economic Adjustment under the Sandinistas* (UNRISD, 1991) and *Economic Reform and Third World Socialism* (Macmillan, 1992).

Trees, people and power

Social dimensions of deforestation and forest protection in Central America.

Peter Utting

UNRISD

EARTHSCAN

Earthscan Publications Ltd, London

To Antonia, Karla, Andrés and the rest of Central America's children

First published in 1993 by
Earthscan Publications Limited
120 Pentonville Road, London N1 9JN

A catalogue record for this book is available from the British Library

ISBN: 1 85383 162 X

Typeset by DP Photosetting, Aylesbury, Bucks

Earthscan Publications Limited is an editorially independent subsidiary of Kogan Page Limited and publishes in association with the International Institute for Environment and Development and the World Wide Fund for Nature.

Contents

List of illustrations *vii*
Acknowledgements *viii*
Introduction *ix*

Part I Deforestation and unsustainable development **1**

1. Central America: Ecological and socio-economic characteristics **1**
Deforestation and environmental degradation 6

2. Causes of deforestation: The processes and players involved **14**
Deforestation and agro-export development 15
Agrarian frontier colonization 17
The cattle boom 19
Logging 21
The expansion of export crops 23
Infrastructural development 27
Fuelwood and urbanization 28

3. Structural and policy determinants of deforestation **34**
Agrarian structure and land tenure 34
Government policy and legislation 37
Deforestation in the 1980s and 1990s: War, agrarian reform, recession and adjustment 41

Part II The breakdown of traditional resource management systems **46**

Overview **46**

4. Deforestation and Indian populations **48**
Central America's Indian population 48
State policy and Indian rights 50
Encroachment in tropical rainforest areas 52
Indians, graziers and colonization in Costa Rica 52
Indians and logging companies in Nicaragua 54

5. Deforestation and livelihood in Guatemala's western highlands **57**
The land tenure system and the demise of communal protection mechanisms 60

	The crisis of petty commodity production and subsistence provisioning	71
6.	**Deforestation and shifting peasant agriculture**	**82**
	Deforestation and land colonization in Panama	82
	Migratory agriculture in Nicaragua	84
Part III	**Forest protection and tree planting initiatives**	**89**
Overview		**89**
7.	**The conservationist approach: National parks and reserves**	**92**
	Protected area schemes	93
8.	**Protected area schemes and social conflict in Costa Rica**	**105**
	Two case studies: The Carara biological reserve and the Osa Peninsula	109
9.	**The project approach: Reforestation, sustainable logging, agroforestry and social forestry schemes**	**113**
	Reforestation	113
	Sustainable forest management	116
	Agroforestry and social forestry schemes	118
10.	**Programme and project implementation: Concrete experiences from El Salvador, Guatemala and Honduras**	**131**
	Community nurseries programme in El Salvador	131
	Reforestation and agroforestry projects in Totonicapán, Guatemala	133
	The Honduran social forestry system	134
11.	**Alternative approaches: Revolutionary change and grassroots mobilization**	**147**
	Radical structural change: Agrarian reform and human resettlement in Río San Juan, Nicaragua	147
	Grassroots mobilization	150
	Deforestation and forest protection in Huehuetenango, Guatemala	158
12.	**Social and political dimensions of forest protection**	**160**
	Social and political economy concerns	160
	External interventions	161
	Social forces and local level structures	167
	Grassroots organization and mobilization	169
Annex 1	Case studies, theme papers and researchers	*174*
Notes		*175*
Abbreviations and Acronyms		*180*
Bibliography		*183*
Index		*193*

List of illustrations

TABLES

1.1 Central America: National, urban and rural population, life expectancy and infant mortality rates 4
1.2 Agrarian structure in Central America (late 1980s) 5
1.3 Central America: Selected socio-economic indicators 6
1.4 Central America: National estimates of forest cover and annual deforestation, 1990 9
2.1 Changes in land use: Selected countries and departments (% of total land area) 20
2.2 Sawnwood production in Central America, selected years 21
2.3 Central America: Forest product exports and imports and trade balance by country, 1970–85 22
2.4 Prices of fuelwood, basic food products and the average wage in Nicaragua, 1982–91 32
3.1 Distribution of rural families by size of holding, 1970 34
3.2 Land concentation in the 1970s 35
4.1 Indian population of Central America (late 1980s) 48
4.2 Costa Rica: Land tenure and forested areas in indigenous reserves, 1990 53
5.1 Agrarian structure in Guatemala and Totonicapán, 1979 59
5.2 Costs of production and revenue in a carpentry enterprise, 1989 76
5.3 Relative prices of selected forest products and other items in Totonicipán, 1980–91 79
7.1 Protected areas in Central America, 1990 92
7.2 Protected areas in Costa Rica, 1990 100
7.3 Costa Rica: Debt-for-nature swaps, 1988–mid 1991 102
9.1 Preliminary financial assessment of selected agroforestry associations in Central America 120

MAPS

i.1 Research sites in Central America xii
1.1 Central America: Topographic features 2
1.2 Central America: Macro-ecological zones 3
1.3 Deforestation in Central America: 1950–85 7
1.4 Deforestation in Costa Rica and Panama 8
2.1 Central America: Agro-export zones 24
5.1 Guatemala: Location of Totonicapán and Sololá 58
5.2 Totonicapán: Forest property regimes 61
6.1 Rainfall and topography in Río San Juan, Nicaragua 88

Acknowledgements

Writing this type of book is a difficult undertaking. It involves attempting to bridge the gap between the natural and social sciences and integrate various disciplines. It also involves working with project practitioners and technical specialists who are sometimes wary of social scientists. Moreover, it requires entering into a dialogue with people in local communities who are often suspicious of outsiders. Thanks to the collaboration and support of numerous people it was possible to piece together the jigsaw which makes up this book. No doubt several pieces are still missing and a few fit uncomfortably but I hope readers will grasp the general picture and appreciate the importance of the issues that are raised.

I am particularly indebted to the researchers and specialists who prepared thematic papers and case study reports: Jens Brüggemann, César Castañeda, Manuel Chávez, Juan Carlos Godoy, Alcides Hernández, Emilio Prado, Rubén Pasos, Carlos Reiche, Santiago Ruíz and Ileana Valenzuela.

Many thanks also to the people in the rural communities where fieldwork was undertaken, who shared with us their thoughts and concerns. I would particularly like to thank those who had the courage to speak out against individuals, groups and institutions which threaten their livelihood and environment.

Numerous other people provided valuable insights, information and advice: Gerardo Budowski, Luis Alberto Castañeda, Marcus Colchester, Dean Current, Ramachandra Guha, Jaime Incer, Stanley Heckadon, Evelien Kamminga, Mauricio Leonelli, Steve Mack, Michael Redclift, Eric Ross, Antonio Ruíz, Graham Woodgate – to mention just a few.

I would like to thank, too, all those who facilitated data gathering and fieldwork by providing logistical support: in particular, Carlos Mendoza and the DIGEBOS team in Quetzaltenango, Guatemala; as well as the very efficient staff at CATIE's documentation centre INFORAT.

My colleagues at the United Nations Research Institute for Social Development (UNRISD) have been most helpful. Special thanks to Solon Barraclough for his guidance and constructive criticism; Krishna Ghimire for providing detailed comments on various drafts; Wendy Salvo for her administrative support; Josephine Grin-Yates, Radhika Jha and Frédéric Grare for their editorial assistance; and UNRISD Director Dharam Ghai, for somehow finding the time to actually read the manuscript and make useful comments.

Finally I would like to thank both UNRISD and the Swedish Agency for Research Cooperation with Developing Countries (SAREC) for providing financial support.

Introduction

During the past decade, the issue of deforestation in tropical countries has captured the attention of the world. Yet despite increasing awareness of range of goods and services provided by forests,[1] no substantive agreement on regulating the use of the world's tropical forests was reached at the Earth Summit held in Rio de Janeiro in June 1992.

Many Third World governments resented the implication inherent in 'the northern position' that they should restrict the dynamic of development centred on the conversion of forests into agricultural land and the growth of forest industries. Given the relative scarcity of capital, skills and technology in the less developed nations, it was to be expected that many would resist calls to curb the use of one of the few 'abundant' resources available to them. This they might do if the price was right but the North was unprepared to pay that price. This was despite the fact that the destruction of tropical forests is also likely to affect the health and future prosperity of the industrialized countries through, for example, carbon emissions and possible global warming, as well as the effects of the loss of biodiversity in the forests, and their status as a store of genetic wealth.

Unfortunately, the debate on the future of the world's tropical forests became bogged down with issues of national sovereignty and economic growth. Important aspects associated with the rights and livelihood of people affected by deforestation processes and forest protection schemes came a poor second or were ignored altogether.

This book is about the people and interest groups with a stake in the forests of Central America. For several decades now, this region has experienced one of the highest rates of deforestation in the world. Current FAO data on deforestation in 12 regions of Africa, Asia and Latin America indicate that only in West Africa was forest cover being removed at a faster rate (FAO, 1991). Here we identify the players, processes and policies that are causing deforestation, examine how the lives of people living in or close to forests are affected when rapid deforestation occurs, look at what is being done in the field of forest protection and tree planting, and consider how social and political factors affect the feasibility of such schemes.

THE RESEARCH FOCUS

The study on which this book is based formed part of a research programme on The Social Dynamics of Deforestation in Developing Countries co-ordinated by the United Nations Research Institute for Social Development

(UNRISD). Under this programme case studies were undertaken in four countries or regions of the world, namely, Brazil, Central America, Nepal, and Tanzania. The research aimed to fill a gap in the existing body of knowledge on environmental problems related to two broad questions: how social forces and government policies underpin deforestation; and how people are affected by deforestation in different ecological and socio-economic settings. As stated in the original project proposal: 'There has been little systematic and comparative analysis of the interactions of deforestation processes at local levels with the associated changes in livelihoods of different social groups . . .' (Barraclough and Ghimire, 1990:2).

The Central American project examined not only the linkages between deforestation and livelihood but also the social impact of forest protection and tree planting initiatives.[2] It soon became apparent on commencing the study that the dynamic phenomenon of the 1980s and 1990s affecting the region's natural resource base was not simply deforestation but conservation policies, programmes and projects intended to protect forests and promote tree planting. From the perspective of human well-being or 'sustainable development',[3] many of these initiatives were having contradictory effects. For this reason it was decided to analyse, in some depth, different experiences or approaches in the field of forest protection and tree planting.

An implicit concern which guided the research was that both the literature on deforestation and the policies designed to protect forests had focused too narrowly on the situation of natural resources and had marginalized important aspects pertaining to people. Such aspects concern the role of different 'players' or interests in processes of deforestation; how deforestation and forest protection schemes affect different social groups; and how people respond when affected by environmental change. Crucial here is an understanding of how forests and trees are integrated in the livelihood systems of local people; how processes of 'disempowerment' of local communities – or the loss of control over resources and decision-making processes – can lead to the breakdown of resource management systems and environmental degradation; and how social structures and land tenure systems condition people's access to resources and, therefore, the distribution of costs and benefits associated with tree felling and tree planting.

Until we begin to address key social and political economy questions such as these, it will be impossible to design and implement effective policies and schemes to protect forests and promote tree planting. An examination of these issues reveals clearly that there are no easy answers or 'technical fixes' for dealing with processes of environmental degradation. Deforestation is part and parcel of the broader problem of under- or maldevelopment, characterized by certain processes:

- social and economic change that enriches the few and marginalizes the many;
- political systems that exclude the rural and urban poor;
- blinkered government policies or development strategies geared towards short-term economic growth; and

- technocratic solutions to problems involving complex social and political issues.

The dimensions and complexity of the problem should not, however, be an excuse for resignation and despair. As this study indicates, there is a way forward but it is one which involves a more integral strategy of change which, in addition to requiring adequate material and human resources, enhanced 'institutional capacity' and improved consultation and dialogue, must also promote processes of local-level 'empowerment' involving grassroots organization and mobilization, a more coherent macro-policy framework and reforms in North–South relations.

All this, of course, is a tall order and may well mean that 'sustainable development', with its glorified imagery of environmental, social and economic well-being for present and future generations, is not a realistic option. Some sort of 'muddling through' or 'patchy development', however, might be possible in order to stem the tide of environmental decline and ensure that conservation strategies do not contribute to further poverty and inequality.

METHODOLOGICAL ASPECTS

The analysis which follows refers to the experience of six of the region's seven countries: Panama, Costa Rica, Nicaragua, Honduras, El Salvador and Guatemala. But unlike a number of other studies on deforestation in Central America, this book does not attempt to take the reader on a country-by-country tour of the region. Rather, it addresses a series of issues and themes, and uses examples and case studies from different countries to illustrate and substantiate the analysis.

The information is drawn from a wide body of secondary sources published in English and Spanish; interviews conducted by the project co-ordinator with government officials, development specialists, academics and representatives of peasant, indigenous and ecology groups; case studies, involving fieldwork, of specific areas and/or social groups in several countries undertaken by individual researchers or small research teams under the supervision of the project co-ordinator; and two theme papers addressing regional issues prepared specifically for this project by specialists. The studies and reports which formed part of the project, as well as the researchers and specialists involved, are listed in Annex 1.

The research was conducted between September 1990 and July 1991. The case studies were designed to fill certain gaps in the existing body of knowledge about:

- the situation of specific social groups in areas experiencing deforestation or affected by forest protection schemes;
- the responses of local populations when their livelihood and lifestyles were affected by deforestation or forest protection initiatives; and
- the capacity of different types of forest protection policies, programmes and projects both to achieve the goals they set themselves and to contribute to human welfare.

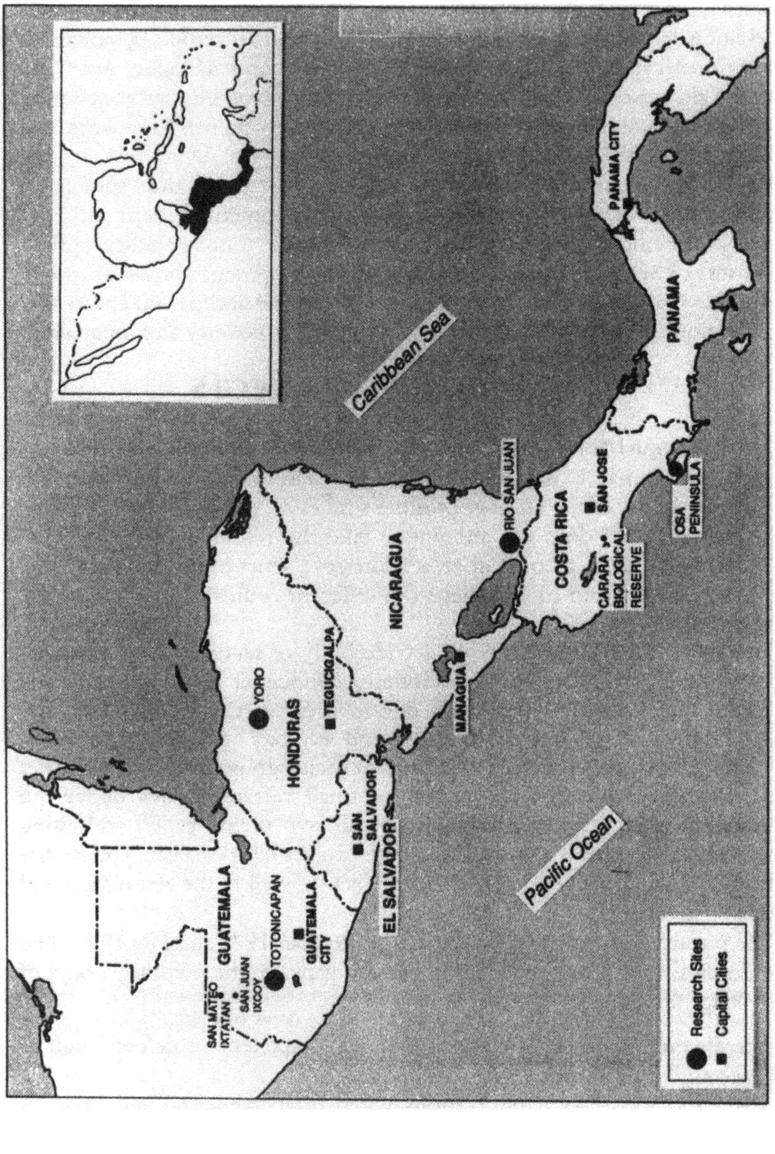

Map i.1 Research sites in Central America

Five case studies were undertaken in Costa Rica, Nicaragua, Honduras and Guatemala. The research sites, indicated in Map i.1, included the departments of Huehuetenango and Totonicapán in Guatemala's western highland or *Altiplano* region, Yoro in Honduras, and Río San Juan in Nicaragua, as well as protected areas in the Osa Peninisula and Carara in Costa Rica. While the depth of the research varied considerably in each case, these studies can best be described as rapid field appraisals. Between one and three researchers would visit a particular area, usually two to three times for several days on each occasion, and interview local peasant or Indian men and women, community leaders, representatives of local producer or community organizations, technicians working in forestry or natural resource protection agencies, and government authorities.

The aim was to begin to grasp the diversity of, and interrelationships between, political economy, socio-cultural, macro- and micro-economic, agro-ecological and biological aspects associated with the social dynamics of deforestation, and the impact and effectiveness of forest protection initiatives. Accordingly, an attempt was made to select experienced researchers or research teams with a multidisciplinary background.

THE STRUCTURE OF THE BOOK

The book is divided into three parts. Part I, consisting of three chapters, examines the causes of deforestation in Central America. Chapter I describes the ecological and socio-economic characteristics of the region and the extent of deforestation and environmental degradation. Chapter 2 identifies the principal pressures which have led to the rapid degradation of forest resources and focuses, in particular, on the way in which so-called modernization processes and 'survival strategies' have led to deforestation during this century. Chapter 3 identifies some of the key institutional and policy determinants of deforestation and examines how contemporary phenomena such as military conflict, agrarian reform, economic recession and stabilization policies have affected the use and management of forest resources. It also considers a number of processes and situations which are specific to individual countries.

Part II analyses the breakdown of certain traditional resource management systems which, historically, were able to provide for a considerable degree of environmental protection and subsistence provisioning. This section identifies the various forces and factors which underpin the breakdown of such systems in recent decades and looks at what happens to livelihood, social relations and social differentiation in local communities when such change occurs. Three specific scenarios of systemic breakdown, involving indigenous and *ladino* peasant groups, are examined with attention focused on the situation of Indian populations in two very different biophysical and social settings.

In this section, Chapter 4 refers briefly to the socio-economic and political situation of the region's Indian population. It then examines how groups in more isolated and sparsely populated rainforest areas of Costa Rica and Nicaragua have been affected by the encroachment of outsiders. Particular attention is focused on how the penetration of lumber companies and cattle graziers served to disrupt the more sustainable land-use practices of Indian groups. Chapter 5 examines the case of Indians living in a specific area of the western highlands or *Altiplano* of Guatemala, characterized by high population density and extreme land scarcity. Despite intense population pressure, communities in this area were, until fairly recently, able to protect their forest resources. During the past two decades, however, this system of communal forest protection has begun to break down. Based to a large extent on a case study report prepared specifically for this project by Ileana Valenzuela, this chapter analyses the factors or pressures which account for this situation and how the lives of local people have been affected.

The breakdown of land management systems practised by *ladino* peasant families,[4] based on shifting agriculture and long crop–fallow rotations, is examined in Chapter 6. Here I draw on a study by Heckadon of colonization processes in Panama and refer also to the situation of agriculturalists in two regions of Nicaragua.

Part III looks at various initiatives that have been taken in the field of forest protection and tree planting. Here I examine with a somewhat critical eye the experience of mainstream approaches generally adopted by governmental, bilateral and multilateral development agencies. Chapter 7 examines the so-called 'conservationist' approach involving the creation of national parks and reserves; while Chapter 8 takes a closer look at the social contradictions associated with protected-area schemes. Drawing heavily on a case study report prepared by Jens Brüggemann, it examines the types of tensions and conflicts which have arisen in Costa Rica when protected areas have been created. Chapter 9 examines the potential and limitations of the 'project' approach involving the implementation of numerous, relatively small-scale, reforestation, 'sustainable logging', agroforestry and social forestry schemes. Chapter 10 goes on to consider, in more detail, the concrete experience of a number of specific programmes and projects involving reforestation and social forestry schemes. These occur in El Salvador, Guatemala and Honduras.

Chapter 11 examines two 'alternative' approaches to forest protection and tree planting involving radical structural change and grassroots initiatives. The former concerns the post-revolutionary experience of Sandinista Nicaragua during the 1980s. At this time, major structural reforms involving population resettlement, agrarian reform and other redistributive policies had a dramatic impact on colonization and deforestation processes in agrarian frontier regions. The second approach involves forms of grassroots organization, mobilization and protest to defend the natural resource base on which local populations depend for their livelihood.

A concluding chapter summarizes some of the lessons which can be learnt from the preceding discussion. It draws attention to the limits to external interventions by governmental or development agencies when planners and policymakers fail to link environmental policies and programmes with the broader development context. Attention is focused on the need to address livelihood issues at the local level, the question of policy coherency at the level of national planning and also reforms in North-South relations. Chapter

12 ends by stressing the crucial role of grassroots mobilization, empowerment and participation in any strategy for sustainable development.

PART I
DEFORESTATION AND UNSUSTAINABLE DEVELOPMENT

Chapter 1

Central America: Ecological and socio-economic characteristics

The Central American isthmus comprises seven countries with a total land area of half a million square km (roughly the size of Spain or Thailand). The region has been described as 'little more than a mountainous barrier between two huge oceanic regimes' (Leonard, 1987:xvii) (see Map 1.1). Broadly speaking, the area can be sub-divided into three major ecological zones (see Map 1.2) (Jones, 1985;1988:243–4; Leonard, 1987:6–8).

The lowland areas of northern Guatemala and those that border the Caribbean Sea form one distinct zone. It is here that the bulk of the region's remaining dense forest cover is located, much of it tropical rainforest. Rainfall varies from approximately 2,000 to 6,000 mm a year and the dry season, where it exists at all, is extremely short.

The range of mountains and valleys which runs through much of the central part of the isthmus constitutes a second climatic zone. This area is particularly extensive in the northern half of the isthmus in Guatemala and Honduras. The climate is more temperate and it is here that most of the region's pine forests are located although small areas of broadleaf or 'cloud forest' – where precipitation is derived from cloud or mist (Synnott, 1989:86) – also exist in some higher reaches.

The Pacific lowlands constitute a third climatic zone. Much of this area experiences virtual drought conditions for between six to eight months of the year. Average annual rainfall generally ranges from 1,000 mm to 1,500 mm and is concentrated between the months of May and October. While most of the forests in this area have been converted to crop and pasture land, small remnants of dry forest and coastal stretches of mangroves exist in certain countries.[1]

Given the region's geographical setting as a land bridge between North and

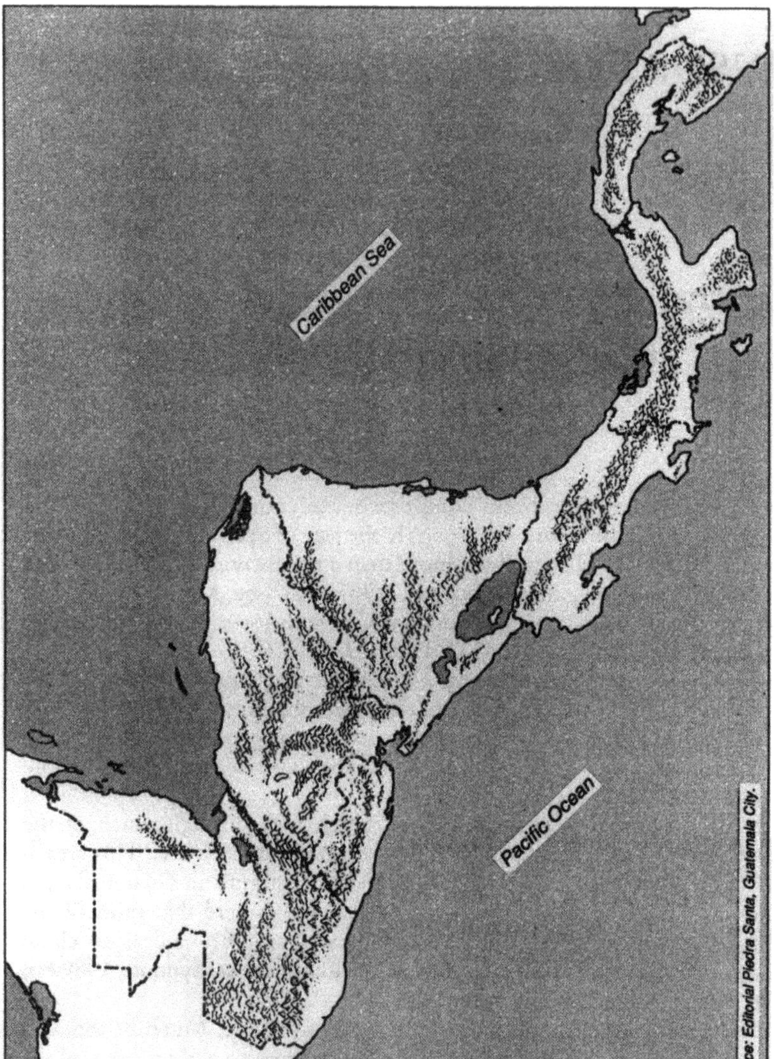

Map 1.1 Central America: Topographic Features

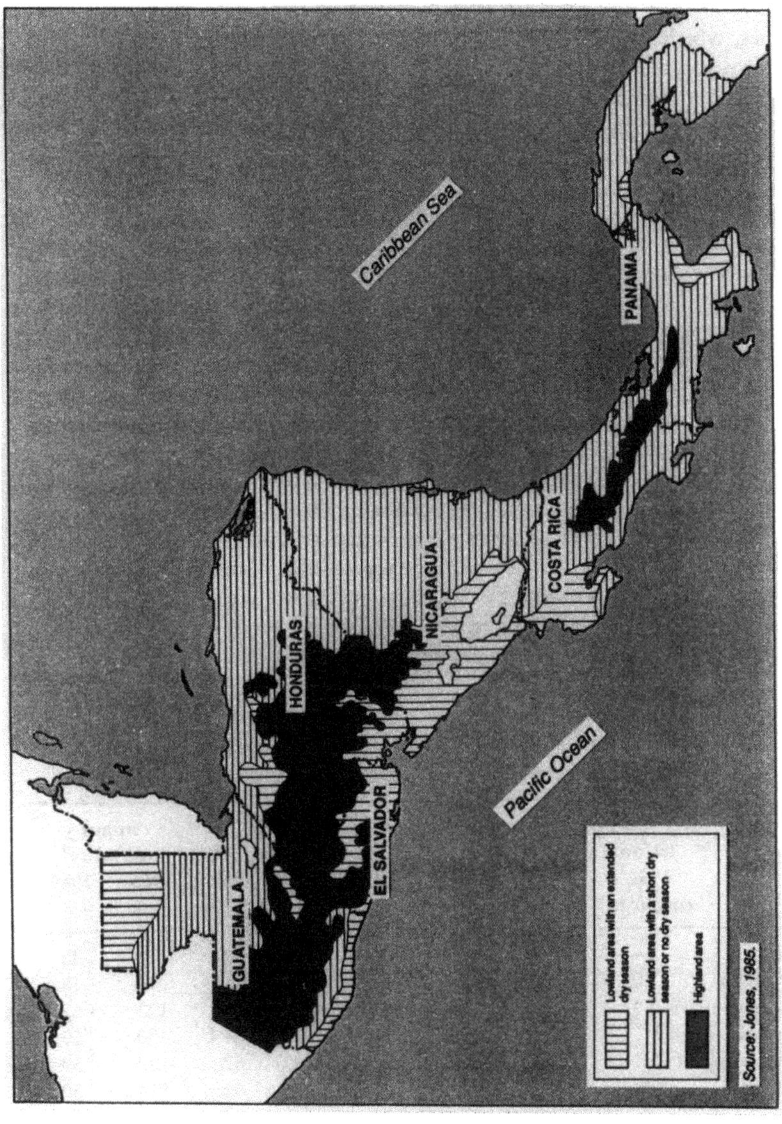

Map 1.2 Central America: Macro-Ecological Zones

South America it is one of the world's most biologically diverse regions (Leonard, 1987). Honduras has at least 700 species of bird, 112 mammals and 196 reptiles and amphibians. The wildlife of Guatemala includes 600 species of bird, 250 mammals and 200 reptiles and amphibians. Costa Rica is known to have at least 848 species of bird while Panama heads the list with 880 species, which is more than all the bird species found in the western hemisphere north of the Tropic of Cancer (Godoy, 1990). Guatemala and Costa Rica are thought to have the most diversified plant life in the region. Costa Rica has approximately 2,000 species of broadleaf trees and 12,000 species of plants. Guatemala has 16 genera (groups of related species) of coniferous trees (ibid.).

The vast majority of Central America's 30 million inhabitants live on the Pacific side or in the hilly interior. The region's population is expanding rapidly. Demographic growth rates vary from 2.1 per cent in Panama to 3.4 per cent in Nicaragua and average 2.8 per cent for the region as a whole. Improved access to health services in recent decades has contributed to a significant reduction in levels of infant mortality and increased life expectancy (see Table 1.1).

Statistics on demographic trends during the 1980s, however, are somewhat unreliable, given the human impact of war, revolution and economic crisis. In Nicaragua, El Salvador and Guatemala, over two million people were killed or uprooted from their homes. Estimates of the numbers of refugees and external migrants vary considerably but the total number of people who fled these three countries is likely to be in the region of 1.5 million. Many of these refugees and migrants settled in Honduras and Costa Rica.

Slightly more than half the region's population lives in rural areas where a highly heterogeneous agrarian society exists. Agrarian structure in Central

Table 1.1 Central America: National, urban and rural population, life expectancy and infant mortality rates

Country	Population			Annual growth 1990[2] (%)	Life expectancy[3]		Infant mortality[4]	
	National 1990[1] (millions)	Rural (%)[2]	Urban (%)[2]		1960	1987	1965	1986
Panama	2.4	45	55	2.1	62	72	58	23
Costa Rica	3.0	46	54	2.6	62	74	72	18
Nicaragua	3.9	40	60	3.4	47	63	123	62
Honduras	5.1	56	44	3.2	46	64	130	69
El Salvador	5.3	56	44	3.2	50	62	122	59
Guatemala	9.2	58	42	2.9	47	62	114	59

Sources: 1. Based on Merino and Mata, 1991.
2. IICA/FLACSO, 1991, based on CEPAL and SIECA.
3. World Bank, 1989. Life expectancy at birth (years).
4. World Bank, 1989. Per thousand live births.

Table 1.2 Agrarian structure in Central America (late 1980s)

Sector	% of total farm area	% of total number of farms
I *Latifundio – Minifundio*	(52)	(54)
– Extensive ranching	46	10
– Subsistence peasantry	6	44
II Agrarian frontier	17	7
III Modernized sector	17	14
IV Small commercial farmers	14	25
Total	100	100

Source: Based on Baumeister, 1991.

America has been characterized in terms of four broad categories of producers or settings (Baumeister, 1991). First, there is the *latifundio/minifundio* complex consisting primarily of large extensive cattle farms and peasant grain producers. While relatively few in number, large cattle ranches account for nearly half the farm land. In contrast, small peasant holdings account for nearly half the total number of farms but only 6 per cent of the land. Second, there is a sector of farmers in agrarian frontier regions who account for approximately 17 per cent of the land and produce mainly beans, coffee and cattle. Roughly the same area of farm land is held by a relatively modern farming sector consisting of relatively large capital intensive holdings producing for both export and internal markets. Finally, there is a sector of small commercial farmers producing coffee, vegetables, sesame, cacao and certain 'non-traditional' crops (see Table 1.2).

International agencies, such as the World Bank, classify the Central American republics as 'lower-middle' or 'middle-income' countries. Annual per capita Gross National Product (GNP) ranged from approximately $750 in Nicaragua (author's calculation based on UNDP, 1991 and CEPAL, 1991) to $1,900 in Costa Rica (World Bank, 1992). Such figures conceal both the extremes of income and wealth which characterize most countries in the region, and the extent of mass poverty. It has been estimated that half the region's population lived in a state of 'extreme poverty' in 1985 (IICA/FLACSO, 1991: Table 1.10). Poverty levels are particularly high in urban areas of Guatemala and El Salvador and in rural areas in Guatemala, El Salvador, Nicaragua and Honduras. Only in Costa Rica are poverty levels relatively low.

Economic crisis and stabilization policies, as well as the impact of war, have had the effect of increasing poverty levels in virtually all countries during the past decade. In constant dollar terms, Gross Domestic Product (GDP) per capita fell in all six countries between 1980 and 1989. Per capita food production levels also fell throughout the 1980s (see Table 1.3).

Table 1.3 Central America: Selected socio-economic indicators

Country	**GNP per capita ($) 1989**[1]	**% living in extreme poverty mid-1980s**[3]		**Per capita GDP**[4]		**Per capita food production index**[6]	
		Urban	**Rural**	**1980**	**1989**[5]	**1979/81**	**88/89**
Panama	1760	16	30	1766	1463	100	90
Costa Rica	1780	10	21	1552	1457	100	91
Nicaragua	830[2]	23	57	747	500	100	58
Honduras	900	32	73	682	601	100	83
El Salvador	1070	40	60	729	521	100	97
Guatemala	910	55	70	1128	924	100	91

Sources: 1. UNDP *Human Development Report* 1992, Table 17.
2. 1988 figure.
3. Percentage of urban and rural population living in extreme poverty (IICA/FLACSO, 1991, Tables 1.14 and 1.18 based on ECLAC).
4. In constant 1980 dollars (IICA/FLACSO, 1991, Table 1.1.3., based on ECLAC and SIECA).
5. Preliminary estimate.
6. World Bank, *World Development Report 1992*, Table 4.

DEFORESTATION AND ENVIRONMENTAL DEGRADATION

During the past three decades, Central America has experienced one of the highest rates of deforestation in the world. Since 1960, the extent of forest cover in the region has been reduced from approximately 60 per cent to just a third of the total land area (see Maps 1.3 and 1.4). Estimates of the extent of forest cover in 1990, based on well-known studies in each country, are presented in Table 1.4. These data generally refer to both closed and degraded forest cover. They indicate that Central America is currently losing about a third of a million hectares, or nearly 2 per cent of its forest cover a year.

Data on deforestation, however, are fairly imprecise. Even at the country level it is normal to find widely varying estimates. This may be accounted for partly by variations in definitions regarding what constitutes forest area but is essentially due to the lack of reliable data. The data problem is brought out clearly in the case of Costa Rica. It was not until 1991 that a study was undertaken which used satellite imagery to compare national forest cover for two years (1979 and 1990). Estimates of deforestation in Costa Rica often range from 30,000 to 60,000 ha per annum. To the disbelief of many national forestry officials, the World Resources Institute (WRI) was still quoting a figure of 124,000 ha in its 1990–91 biennial report (*Tico Times*, May 31, 1990:24). Two major reports referring to deforestation in Guatemala, both published in 1990, cite figures of annual forest loss ranging from 40,000 to 90,000 ha (Guatemalan government, 1990; Bradley *et al.*, 1990b, respectively).

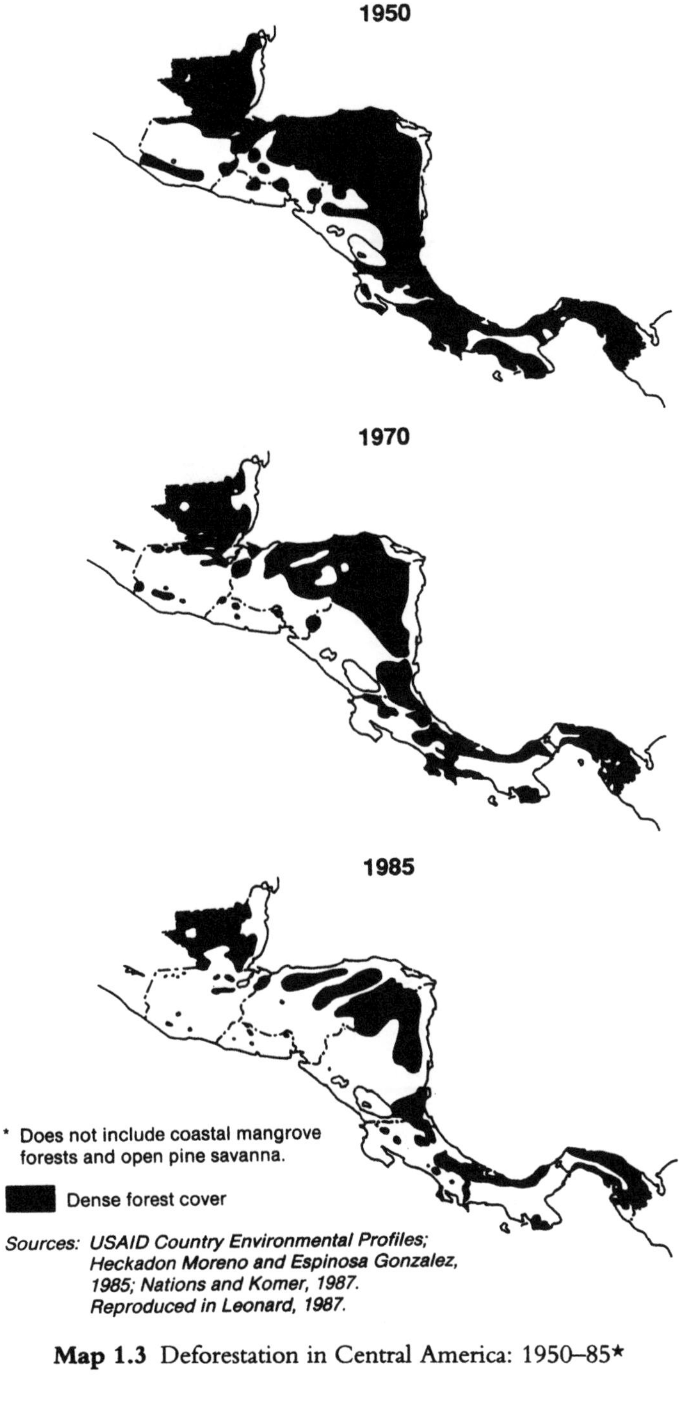

Map 1.3 Deforestation in Central America: 1950–85*

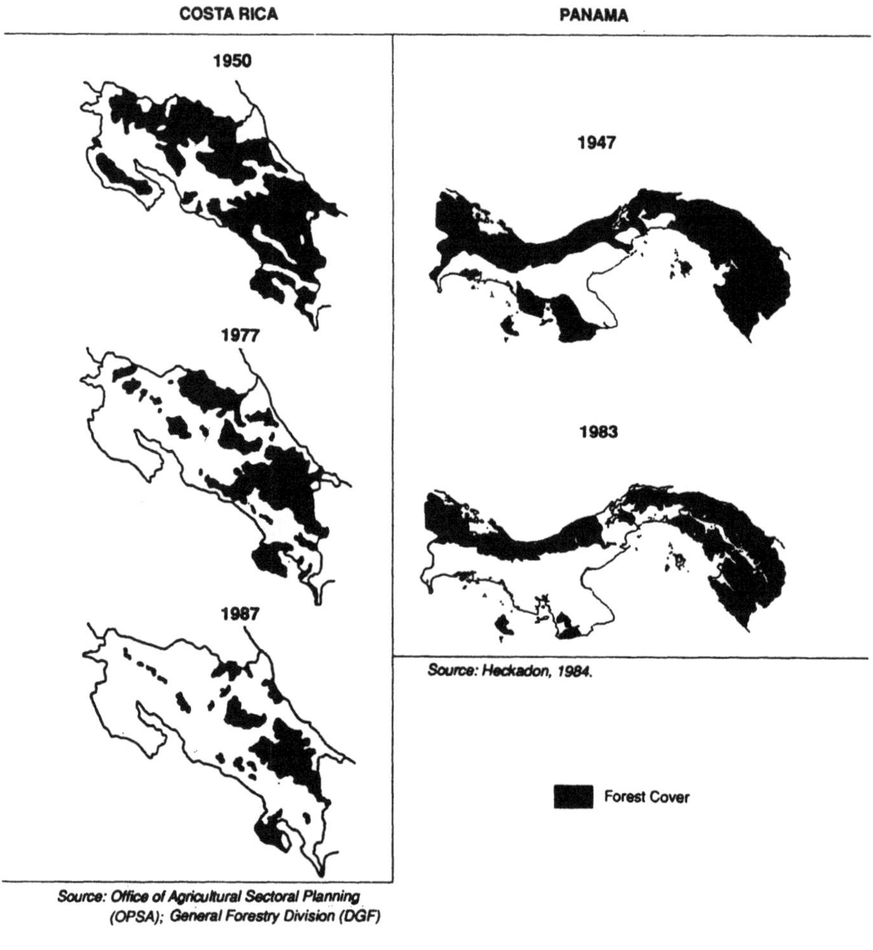

Map 1.4 Deforestation in Costa Rica and Panama

Table 1.4 Central America: National estimates of forest cover and annual deforestation, 1990

Country	Forest area (000 ha)	% of total land area	Annual deforestation (ha)
Panama	3,203[1]	42	34,000[7]
Costa Rica	1,426[2]	33	50,000[8]
Nicaragua	4,140[3]	30	70,000[9]
Honduras	4,731[4]	42	80,000[10]
El Salvador	250[5]	12	14,000[11]
Guatemala	3,762[6]	35	90,000[12]
Total	17,502	36	338,000

Sources: 1. Estimate for 1990 based on 1987 figure of 3,305,300 ha cited in INRENARE, 1991:13.
2. Estimate for 1990 based on 1989 figure of 1,475,940 ha cited in MIRENEM, 1990.
3. Refers to closed and degraded forest. Ministerio de Agricultura y Ganadería (MAG), 1990.
4. Estimate for 1990 based on COHDEFOR (SECPLAN *et al.*, 1990).
5. Mansur, 1990. Figures refer to closed and degraded forest.
6. Guatemalan government, 1990.
7. INRENARE, 1991:13.
8. Bonilla, 1988:72; MIRENEM, 1990:4.
9. MAG, 1990.
10. SECPLAN *et al.*, 1990.
11. Mansur, 1990.
12. Bradley *et al.*, 1990b.

While estimates of forest cover and rates of deforestation may vary considerably, it is clear that the rate of deforestation has increased sharply during the second half of this century. Only in El Salvador did what could be described as an ecological crisis exist prior to this period. It has been estimated that by 1900 only 10 per cent of the original forest cover remained (USAID, 1985:5). Following a tour of the country in 1946, the conservationist William Vogt identified serious problems of erosion and declining agricultural productivity, the increasing incidence of flooding caused by silting of rivers, semi-desertification in areas such as Morazán, and the 'reckless sacrifice' of flora and fauna:

> The destruction of the lands of El Salvador is almost universal wherever there are slopes – and the country has but little level land – a very high proportion of the land now being farmed in El Salvador is slowly and – under current methods of treatment – inevitably marching toward the sea. (Vogt, 1946)

The seriousness of the contemporary ecological situation in much of the region is reflected in facts and figures gathered from reports published in each of the Central American countries.

Panama

Panama is considered to be one of the Latin American countries most affected by erosion. One FAO report estimates that some 90 per cent of the total land area was affected by erosion by the mid-1980s while 16 per cent was seriously affected (FAO, 1986:48–9). The area under forest cover has been reduced from 70 per cent in 1947 to just over 40 per cent in 1990 (INRENARE, 1991). The future existence of the country's economic lifeline – the Panama Canal – is seriously threatened by sedimentation caused largely by deforestation in the upland areas of the canal zone watersheds (Alvarado, 1985; Rubinoff, 1982; Wadsworth, 1982).

Costa Rica

In Costa Rica the percentage of total land area covered by dense forest was reduced from approximately 75 per cent to 20 per cent between 1940 and the mid-1980s (Silliman, 1981; Chacón *et al.*, 1990). The exploitable forest area remaining outside of national parks and reserves amounts to less than a quarter of a million hectares or 5 per cent of the national territory (Hedström, 1990; MIRENEM, 1990). It has been estimated that the commercial supply of forest resources could be exhausted by the year 2003 if current trends persist (Arcia *et al.*, 1991) or even earlier (mid-1990s), according to other sources (Finegan and Saboval, 1989).

Nicaragua

During the 1960s and 1970s, Nicaragua experienced the highest rate of deforestation in the region, losing each year approximately 100,000 ha of forest. While the rate of deforestation declined during the war years of the 1980s, it has increased sharply in recent years as migrants return to agrarian frontier regions. Also, in October 1988, Hurricane Joan destroyed or partially damaged 560,000 ha of dense rainforest, 17 per cent of the country's total forested area. Deforestation has resulted in major environmental damage on the Pacific side of the country, causing serious erosion, flooding and changes in micro-climate (Larson, 1989; MAG, 1990; Peters, 1985).

Honduras

Between 1964 and 1986 Honduras lost a quarter of its forest area. Average annual deforestation was of the order of 80,000 ha (SECPLAN *et al.*, 1990). Fires in pine forest areas in 1987 and 1988 affected an average of 95,000 ha each year and the regeneration of the country's pine forests has been seriously affected by cattle grazing. The environmental consequences of deforestation have contributed to serious water shortages in the capital city, Tegucigalpa.

El Salvador

El Salvador is generally regarded as the most environmentally deteriorated country in continental Latin America. Only 240,000 ha of largely degraded forest remain[2] while just 3 per cent of the country's original forest cover still exists. Recent years have seen the destruction of mangroves along the Pacific coast, reduced from an estimated 120,000 ha to 30,000 ha. At current rates of deforestation the country's entire stock of forest resources will be depleted by the year 2005 (Mansur, 1990). Erosion affects an estimated 77 per cent of the national territory (USAID, 1985; CESTA). Every major watershed and river basin is degraded. Floods and landslides are commonplace and sedimentation threatens the country's hydroelectric system (Hall and Faber, 1989:4).

Guatemala

In 1960, 77 per cent of Guatemala was under forest cover; by 1980, only 42 per cent remained (Bradley *et al.*, 1990b) The country currently experiences one of the highest rates of deforestation in the region with some estimates putting the annual loss of forest area as high as 90,000 ha (ibid). Since 1960, over half of the country's 500 square km of mangroves have been destroyed (Godoy, 1992:88) An estimated one-third of Guatemala's land mass is seriously eroded or degraded.[3]

There is a considerable body of literature documenting the environmental impact of deforestation in Central America. Data on erosion presented above indicate that Panama, El Salvador and Guatemala probably rank amongst the world's most eroded countries. Soils in much of the region are fragile in two important respects.

First, there are the fertile volcanic soils of certain Pacific coastal and hilly interior regions of Guatemala, El Salvador, Honduras and Nicaragua which are highly prone to erosion once forest cover is removed (Leonard, 1987:13–14). Not only the removal of trees but also the annual burning of vegetation leaves vast areas bare at the onset of the rainy season and vulnerable to erosive effects of wind and rain.

Secondly, there are the soils in northern Guatemala and in much of the Caribbean coastal strip which runs the entire length of the region. Here the fertile topsoil layer is generally quite thin and soils are prone to leaching. Agriculture and deforestation in many such areas are associated with serious environmental problems (Leonard, 1987: 13–14; Brüggemann and Salas, 1992).

In both regions, problems of soil fertility and erosion have also resulted from soil compactation due to the movement of cattle. Compactation tends to lead to increased water run-off as the number of large and medium soil pores is reduced (Brüggemann and Salas, 1992). Increased erosion and aridity or drought often ensue, particularly in hilly areas with an extended dry season. In both wet and dry regions, compactation can lead to changes in the organic content of the soil and reduced soil fertility (ibid.).

Soil fertility, notably in pasture areas, has also declined as a result of the destruction of the nutrient cycle associated with micro-organisms in the soil. A study of environmental problems in the Puriscal region of Costa Rica describes this process in the following terms:

> The practically irreversible loss of biodiversity appears to be another major problem of the pastoral system in the region. According to reforestation research in old pasture areas, it seems to be extremely difficult to reforest these areas with native species without special application of humus and labour. These soils under pastures have lost a large part of the micro-organisms contributing to plant nutrition, that is, the Mycorrhizae (a fungus which lives in symbiosis with the plant's roots). It seems that because of the lack of dense vegetation during the dry

> season, the micro-organisms could not survive the drying processes of the soils. (Brüggemann and Salas, 1992)

As these authors point out, such problems are likely to be more accentuated both in areas with poorer quality soils and in dry areas where burning generally has a much more damaging effect on the root structure.

The socio-economic and agro-ecological effects of these processes of environmental degradation can be dramatic. In several countries, areas which were once considered national bread baskets have become virtual deserts. Referring to Panama, Rubinoff writes:

> The Azuero Peninsula was once called the bread basket of Panama. But the trees were cut down, initially to grow maize, then to raise cattle.... Today the Azuero Peninsula is rapidly becoming a desert and the inhabitants are migrating elsewhere. (Rubinoff, 1982:3)

Referring to Costa Rica, Silliman writes:

> Such extensive changes in the natural vegetation are not without broad environmental consequences. Erosion, falling river levels, flooding, and unproductive farmlands have already become serious problems in many areas.... A typical example is the Parrita watershed, just south of San José. Once the bread basket zone of Costa Rica, it produced the majority of corn and beans in the country. Now it is entirely devoted to pasture, due to a decline in soil fertility. Landslides and road washouts are common in the entire zone. The force of tropical rainstorms, once dissipated by the natural forest canopy, now strikes the open ground directly, increasing erosion. Instead of percolating into the soil to be retained by the root structure of trees, rainwater now runs directly off the surface, causing floods in the banana plantations downslope. Large areas of soils have been so severely impoverished and eroded that they no longer even yield pasture, and some experts doubt that they can ever again support forests. This problem is now widespread along the entire central and south Pacific slope. (Silliman, 1981:66)

As indicated earlier, the ecological crisis in El Salvador is fairly generalized throughout the country. The situation has been described by Hall and Faber in the following terms:

> Much of the northern mountains are already irreversibly damaged, stripped of their topsoil, with only scrub and sparse undergrowth, or bare rock, left behind. Groundwater is disappearing, and many *campesinos* find their wells drying up. Every major watershed and river basin is degraded, including the Río Lempa, which stretches across the country and is the single largest watershed in all Central America. Floods and landslides in the rainy season and droughts in the dry season cause millions of dollars of losses to the Salvadorean economy, while sedimentation deposits from soil erosion threaten to cripple the dams of the country's hydroelectric system. (Hall and Faber, 1989:4)

Erosion and declining yields pose a serious threat to peasant agriculture throughout much of Central America. This is particularly the case in hilly and mountainous areas towards which peasant production has been pushed primarily by commercial agriculture. The crisis of peasant agriculture in such a context has been described in various studies. According to a USAID report on Honduras:

> Deforestation by the shifting cultivator clearing high forest in Yoro and Olancho is dramatic and well-publicized. The human tragedy is even more serious for the many thousands of *campesino* families living on degrading lands in the Choloteca Valley and the western departments bordering El Salvador. The forest cover has been peeled back leaving a threadbare patchwork of grazed bush fallow and cultivated plots.
>
> Increased population pressure and the resultant shortened fallow period interrupt the process of regeneration in soil fertility, introduce erosion hazard, and result in energy inefficiency since soil productivity declines.
>
> Given existing technology, this mountainous land is not likely to be a major source of products for national and export markets. Therefore, the tendency is to abandon this land and its *campesinos*, and invest existing resources in improving agricultural production among more sophisticated farmers on the better valley soils. Such a course of action will have serious social, economic and environmental effects. (Campanella *et al.*, 1982:8)

While there is ample evidence pointing to serious environmental degradation throughout the region, there is also a tendency to overstate the extent of the 'crisis'. Much of the literature on Costa Rica, for example, conjures up images of 'paradise on the brink',[4] partly due to rapid deforestation. But a number of local-level studies indicate that the situation may not be quite so critical as is often portrayed. One study of erosion in the Costa Rican canton of Puriscal, which was declared an 'Emergency Zone' by the government during the late 1970s due to the perceived environmental crisis, found erosion levels to be considerably lower than expected (Vahrson and Cervantes, 1991). While declining soil fertility and erosion in the area constitute serious problems, it is doubtful that they were primarily responsible for the change in land-use patterns which caused the Parrita watershed to lose its status as the country's bread basket (as indicated in an above quote). This was due largely to economic factors associated with changes in relative prices and incentives for cattle raising, as well as processes of land concentration (Brüggemann and Salas, 1992).

Another crisis scenario often presented in the literature is that associated with the destruction of rainforests in Costa Rica's Atlantic coast region. In parts of this area, however, relatively fertile soils do exist, due to their association with volcanic activity in the central highlands. As one soils specialist based in the northern Atlantic area explained: 'Here our worst soils would be about the best in Brazil's Amazonian region'.[5]

These final comments are not intended to deny that major environmental problems exist or that they are getting worse. They merely serve to draw attention to the facts that assumptions are often made on the basis of flimsy evidence and hearsay; that environmental problems that characterize a specific ecological zone are often assumed to exist in zones with very different characteristics; and that there is sometimes a tendency to underestimate the suitability or sustainability of certain farming systems which function and evolve once areas have been cleared of forest.

Chapter 2

Causes of deforestation: The processes and players involved

When reading about tropical deforestation, be it in Central America or in other Third World regions, one is usually presented with lists of causes which tend to highlight the roles played by shifting peasant agriculture, the conversion of forest to pasture, the expansion of certain agro-export crops, uncontrolled logging activities, fuelwood gathering and urbanization. Rapid population growth is often posited as one of the major forces resulting in increased pressure on the natural resource base which, in turn, results in deforestation.

While most authors generally agree on which agents are directly involved in cutting down trees, they often disagree in their assessment of what are the primary causes, the appropriateness of land-use patterns associated with deforestation, the underlying social forces which determine such patterns, and the question of who, if anyone, is to blame. Increasing pressure on the land, for example, is clearly a fundamental force underlying deforestation but is it population growth or skewed patterns of resource distribution and inappropriate land-use patterns which are responsible? Are peasants at fault because they are clearing and burning large areas of forest or are they the victims of a particular socio-economic system which has made access to land and other resources in areas of greater agricultural potential increasingly difficult? Do government policies which encourage colonization and cattle raising in rainforest areas help or hinder the development process?

These questions are important not just for analytical clarity but because they raise crucial policy issues in what are often very contentious areas – population control, land reform, income redistribution, intensification of land use versus colonization of virgin areas, and so forth. The bottom line in much of the debate regarding deforestation is whether the way forward is along a path of technical and administrative 'fixes' – creation of national reserves, reforestation, sustainable forest management, agroforestry schemes, etc – or whether a more fundamental restructuring of income, wealth and power, at local, national and international levels, is ultimately required.

The debate regarding the role of rapid population growth in deforestation processes pervades much of the literature on this subject. Researchers from the Costa Rican Demographic Association have analysed the linkages between population growth and environmental degradation in some depth. They claim that the population factor and the increased demand for land, food and wood associated with population growth, should, following Shaw (Shaw, 1989), be regarded as 'proximate' causes. These accentuate or intensify the effects of more fundamental or 'ultimate' causes of a socio-economic character such as inegalitarian structures, government policies and cultural aspects (Arcia *et al.*, 1991; Merino and Mata, 1991).

This may indeed be the case; but the analysis tends to avoid important questions concerning the conditions under which increased population does or does not result in deforestation and environmental degradation (or, for that matter, innovation and environmental rehabilitation – see Blaikie and Brookfield, 1987:27–33), of the relative suitability or sustainability of different farming systems which operate once areas have been cleared, and of the relative importance which should be assigned to the population factor as a cause of deforestation and environmental degradation.

Clearly, no simple correlation can be drawn between population pressure and deforestation. As we will see in Chapter 5, one of the rural areas with the highest population density in Central America was, until fairly recently, one of the best-protected forest areas in the region – even without the backing of government legislation to control tree felling. El Salvador is often held up as a classic example of a country where the destruction of the natural resource base has been caused by rapid population growth. During this century El Salvador's population increased from less than a million to over 5 million and the country became the most densely populated nation on the Latin American continent. As indicated below, however, other processes have played a crucial role.

What emerges from the analysis that follows is that there are no simple, uni-causal explanations of deforestation; that the various specific 'causes' are part and parcel of a particular style of development or development strategy which promoted, simultaneously, agro-export expansion and social exclusion; that many causes are closely interrelated; and that blame is often apportioned incorrectly.

DEFORESTATION AND AGRO-EXPORT DEVELOPMENT

The 'causes' of deforestation referred to above are all features of a specific style of development which unfolded during the late 1800s and took off during the latter half of this century. They are, in fact, related to certain processes of economic and social change which characterize the so-called agro-export model. These may be summed up in three words: 'marketization', modernization and marginalization.

'Marketization'

'Marketization' or market integration, operated at two levels. First, national economies were drawn into the world market for primary goods – notably coffee, cotton, sugar, bananas and beef. In several countries, this process began during the latter half of the last century when world demand for Central American coffee rose sharply. The scale of deforestation and environmental deterioration associated with coffee expansion, however, was not as intense as that which occurred from the 1950s onwards when the countries of the region experienced booms in several agro-export product sectors.

Secondly, peasant producers were drawn increasingly into national markets. The intensification of commodity relations meant that rural families not only had to produce much of the food they required but also a marketable surplus in order to obtain the income necessary to purchase production inputs and consumer goods and services. This led to an intensification of agricultural production which in many areas of Central America broke the fragile ecological equilibrium that characterized traditional slash and burn agriculture, whereby land had to be left fallow for many years before it could be cultivated for relatively short periods. It also accelerated the conversion of forest areas to crop and pasture land.

Modernization

Modernization implied a certain vision of progress and set of cultural attitudes – generally adhered to by development planners and policymakers, banks, logging companies, graziers and other commercial farmers – which saw the forest as an obstacle to development. This view contrasted sharply with that which, historically, characterized Mayan and several other indigenous cultures of the region. Cultural perceptions, however, are in a state of flux. The process of acculturation affecting Indian groups has considerably weakened traditional beliefs regarding the sanctity of the natural elements. It has also weakened the extremely rich knowledge base which underpinned the sustainable use of forest resources and enabled indigenous groups to use numerous forest resources for basic needs provisioning.

Modernization involved the rapid conversion of forest to crop or pasture land in order to pave the way for short-term increases in production with little regard for the long-term sustainability of production systems. It also involved the introduction of new technologies, some of which, being inappropriate for the type of ecological conditions prevailing in tropical forest areas, resulted in greater risk for the small producer.

Modernization also implied the development of economic infrastructure such as roads, railways and hydroelectric power schemes which directly or indirectly accelerated deforestation. It meant, for certain groups and geographical areas, improved health care facilities and reduced levels of infant mortality which in many rural areas resulted in larger families to feed, clothe and care for, and hence increased demands on the land and agricultural production.

Marginalization

Marginalization refers to highly skewed patterns of resource distribution which characterized the agro-export model and that left the mass of the region's population living in poverty. It is also associated with processes of 'disempowerment' which implied reduced rights and control for large social groups over physical and social resources as well as decision-making processes associated with resource management (Vivian, 1991). Poverty and the limited access of the mass of the rural population to land, credit and other essential resources resulted from an intense process of social differentiation associated with agro-export development.

The improved opportunities for commercial farming which accompanied the insertion of the Central American economies in world commodity markets prompted processes of land concentration and the displacement of peasant producers from areas suitable for food production. Interests (both national and foreign) involved in the production, processing, and trading of agro-export products, as well as the financing of such activities, came to exert a dominant influence over national states. Government policies and development programmes generally favoured such groups and often ignored the needs of peasant producers and indigenous groups.

Marginalization also underpinned, to some extent, rapid population growth and related pressures on the land in many rural areas. Having more children acted as 'a hedge against poverty' (Hall and Faber, 1989:5) and, in the absence of capital and intermediate goods (chain saws, agrochemicals, etc.) or money to hire wage labourers, it also provided families with the one resource that was necessary to gain access to land in agrarian frontier regions, namely labour power (Howard, 1987b).

AGRARIAN FRONTIER COLONIZATION

Much of the land which has been deforested in Central America has been cleared either by shifting peasant cultivators or ranchers colonizing agrarian frontier regions. Most countries still possess an agrarian frontier and experience colonization processes which have a direct bearing on the question of deforestation. Such processes can act as an escape valve to relieve pressure on the land in other areas. This is not the case, though, in El Salvador while in Costa Rica the stock of forest land outside of protected areas has virtually disappeared already.

The environmental problematic associated with areas experiencing colonization not only involves the fact that large areas of forest have been destroyed and potentially valuable forest resources have literally gone up in smoke, but also that certain farming systems which have been put in place have proved unsustainable. Part of the problem resides in the 'geographical trap' which characterizes agrarian frontier colonization in much of Central America. Peasants and farmers have been pushed increasingly towards ecologically fragile areas – steeper hillsides, and/or areas with poorer soils and higher rainfall. In such areas, problems of erosion and leaching, as well as plant, animal and human disease often intensify. Moreover, many who colonize rainforest areas bring with them knowledge and farming practices which evolved in a very different agro-ecological context and which often prove inadequate for the conditions that characterize agrarian frontier regions.

The extremely high rates of deforestation experienced in Nicaragua in the 1960s and 1970s resulted primarily from the clearance of forests by graziers and peasants in agrarian frontier areas. The same process in Honduras is thought to have accounted for 80,000 ha of forest each year during the 1970s

(Campanella *et al.*, 1982). Today, the region's main colonization fronts are located in Panama, Nicaragua, Honduras and Guatemala (Jones, 1988:249).

Panama's principal colonization front is the Darién region which borders Colombia. Since the 1950s, this area has attracted migrants from other provinces, notably Chiriqui and the Azuero Peninsula (Hernández, 1982). During the early 1980s, rapid spontaneous colonization also occurred in the western part of the country (Jones, 1988:249).

Colonization of Nicaragua's Atlantic coast region accelerated when the cotton boom of the 1950s and 1960s restricted access to land in the Pacific coastal regions, and when the demand for pasture land increased during the cattle boom of the 1960s and 1970s. The process accelerated during the 1970s when the Somoza government initiated an ambitious colonization project in south-eastern Nicaragua. The initial phase of the 'Rigoberto Cabezas' project, which began in 1972, covered 50,000 ha but formed part of a broader plan covering an area of 400,000 ha (Instituto Agrario de Nicaragua, 1972). As discussed below, the war between the Sandinistas and the US-backed *Contra* rebels put a partial brake on colonization during most of the 1980s. It has accelerated again, however, during the early 1990s following the peace accords, the return to the country of thousands of refugees and exiles, and the intensification of the economic and social crises which have driven people to seek alternative livelihood opportunities.

It is probably in Honduras where the process of colonization of virgin areas was most intense during the 1980s (Jones, 1988:253). This dynamic has been fuelled by three major forces, identified by Jones. First, land concentration and mechanization associated with agro-export development in the south of the country, combined with high population density, forced many to leave the area and migrate. Second, a number of government projects promoting the expansion of agro-export and agro-industrial development have encouraged this process. Thirdly, thousands of refugees from Nicaragua and El Salvador poured into Honduras. Many settled in camps and communities located in or near forest areas.

The colonization process in Guatemala has centred on the country's northern Petén region and a long thin strip of land, stretching across central Guatemala, known as the Franja Transversal del Norte (FTN). During the past decade, colonization in this area has been directed largely by the state, which during the early 1980s drew up plans to establish 'development poles' in 44 remote areas (ibid.:254). The involvement of the military in the colonization process has been considerable. Military officers have been recipients of large tracts of land and have also promoted the development poles as part of a counter-insurgency programme (Chapin, 1987) to establish free-fire zones and disconnect guerrilla groups from their social base of support in rural areas.

While we are often presented with an image of hordes of peasants and ranchers invading agrarian frontier regions, wielding their axes or chain saws to quickly clear vast areas of forest, it is important to adopt a more nuanced vision of the colonization process. This is apparent in at least two respects. First, as explained below in the discussion on logging, colonization has often involved various stages and several different groups of actors. Hence, land-seeking peasants have often followed in the wake of the logging company, or the rancher has followed the peasant or land speculator.

Secondly, rapid and extensive deforestation is not always the immediate result of colonization. It has been noted, for example, in certain areas of Costa Rica's northern Atlantic zone, that the initial colonizers cleared relatively small areas of forest. This was done primarily for the purpose of staking a claim by marking boundaries. Establishing farming systems in the area proved to be extremely difficult and forest clearance for crops and pastures proceeded relatively slowly (Brüggemann and Salas, 1992). As these authors point out, such a situation can have important policy implications. Very often, for example, satellite imagery is used to identify areas appropriate for protected area schemes, on the basis of whether or not areas of rainforest are uninhabited or unclaimed. The extent of forest cover is the crucial indicator. This example indicates, however, that the correlation between deforestation and human presence may be erroneous. Under such conditions the establishment of a protected area is likely to lead to considerable tensions with those with land claims or property titles.

THE CATTLE BOOM

From the mid-1950s to the mid-1970s, the area under pasture in the region increased from 3.9 million to 9.4 million ha, nearly a fifth of the total land area (Williams, 1986; Heckadon, 1984). By the 1980s, approximately two-thirds of all agricultural land was covered by pasture (Heckadon, 1992:14).

Colonization of agrarian frontier regions and rates of deforestation increased sharply in Honduras, Nicaragua and Costa Rica during the 1960s as land was cleared to make way for pasture. This was in response to the rapid growth in United States demand for Central American beef – what has been referred to as the 'hamburger connection' (Nations and Komer, 1987).

When analysing the relationship between cattle expansion and deforestation it is also important to recognize that the preferences of local farmers and ranchers for cattle have to do with much more than straight dollars and cents or export-driven demand. This is particularly so in certain agrarian frontier or tropical rainforest areas where cattle perform a multiplicity of functions (Brooijmans and van Sluys, 1990; Brüggemann and Salas, 1992). In situations where property titles do not exist, cattle can facilitate access to credit by acting as collateral. They also provide security in what is a highly risky environment by functioning as assets that can be sold as and when needed. Cattle enable producers to move and market their produce in situations where roads and mechanized transport are limited. Moreover, cattle ranching is important in cultural terms, providing farmers with a much valued identity and social status.

The trade-off between pasture expansion and forest area is illustrated dramatically by Williams when comparing land-use patterns in the major cattle-producing departments of Guatemala, Honduras and Costa Rica before and after the rapid growth of beef production (see Table 2.1).

Table 2.1 Changes in land use: Selected countries and departments (% of total land area)

Country	Department	Period/year	Pasture (%)	Forest (%)
Guatemala	Escuintla	Pre-export boom	12	48
		1964	42	16
Honduras	Choluteca	Pre-export boom	47	29
		1974	64	11
Costa Rica	Guanacaste	Pre-export boom	39	34
		1974	65	13

Source: Based on Williams, 1986:113.

While different methods were used to convert forest to pasture, it was common for landowners either to buy 'improved land' from shifting cultivators or to allow peasant producers to clear land for basic grain production, farm it for one or two seasons, and leave it ready for seeding when they moved on to another plot. In the words of one writer: 'As the cattle boom progressed life on the edge of the peasant system consisted of being chased toward an ever-vanishing frontier' (Williams, 1986:117).

Those who cleared the land were sometimes land speculators who quickly cut and burned areas of forest and sold the 'improvements' to peasant families or ranchers. This practice was particularly evident in Costa Rica where the area under pasture increased from 12 per cent to 33 per cent of the total farm area between 1950 and 1984, while the percentage share of forest land fell by 35 per cent to 16 per cent (MIRENEM, 1990:4).

The cattle sector's appetite for destroying forests was determined partly by technological factors. Growth in production tended to occur not via increases in productivity but, rather, expanding herds and area (Howard, 1987b). The ratio of head per hectare remained low. One study of Nicaragua indicates that, on average, one *manzana* (0.7 ha.) of pasture was required for every additional head of cattle and that much higher ratios were common on poorer quality land (ibid.).

The increase in demand for pasture area did not always lead immediately to the clearing of primary forest. One analysis of census data on land-use patterns in Honduras, for example, has shown that the considerable expansion of pasture area between 1952 and 1964, was mainly at the expense of fallow land (Howard, 1987a). This process, however, clearly had important implications for the sustainability of traditional peasant farming systems based on crop and fallow rotation (necessary to recuperate soil fertility). It no doubt accelerated trends associated with the crisis of peasant agriculture whereby peasants came to rely increasingly on waged employment and/or migrated to agrarian frontier or urban areas. It was during the late 1960s and throughout the 1970s that the increase in pasture area in Honduras occurred, primarily via the conversion of agrarian frontier forest land (ibid.). The cattle boom, therefore, should be associated not simply with deforestation but also with the breakdown of traditional resource management systems.

The upshot of these trends was that household subsistence provisioning became increasingly difficult for many peasant families who were forced to farm poorer quality land as the area of fallow land declined, and who found it increasingly difficult to find employment in rural areas. Migration to urban centres or the agrarian frontier became the logical option for hundreds of thousands of people throughout the region. As Howard points out, the cycle of cattle expansion, deforestation and migration continued in the agrarian frontier (ibid.).

LOGGING

Colonization was often facilitated by the operations of lumber companies (Budowski, 1990; Nations and Komer, 1987). It is generally the case, however, that the forestry industry of the different Central American countries is not particularly large. Data presented in Table 2.2 indicate that only in Costa Rica, Honduras and Nicaragua does sawnwood production constitute a significant economic activity. Moreover, levels of sawnwood production appear to have fallen in most countries during the 1980s.

Only in Honduras does the lumber industry constitute a major economic sector and foreign-exchange earner. It has been estimated that the operations of the logging companies in Honduras deforest or degrade some 25,000 ha each year (Hernández, 1986:8).

In many forest areas, tree cutting is highly selective. In the broadleaf forests of Honduras, for example, normally two to three species are cut, while trees that are felled in the pine forests tend to be those of the highest genetic quality (SECPLAN *et al.*, 1990:143). Throughout the region's broadleaf forests, it is usually the case that just two or three trees per hectare are removed. In Honduras, the average for species such as mahogany is even less.[1] As Budowski explains:

Table 2.2 Sawnwood production in Central America, selected years (thousands of cubic metres)

Country	**1953**	**1965**	**1970**	**1975**	**1980**	**1988**
Panama	22	71	44	50	53	45
Costa Rica	207	308	400	612	524	515
Nicaragua	130	107	195	402	402	222
Honduras	310	655	449	551	546	437
El Salvador	–	20	20	38	37	44
Guatemala	173	164	72	222	93	83
Total	842	1325	1180	1875	1655	1346

Source: United Nations *Statistical Yearbook*, (1970, 1975, 1983/84), New York.

Table 2.3 Central America: Forest product exports and imports and trade balance by country, 1970–85 (millions of US dollars)

Country	Exports 1970	Exports 1985	Imports 1970	Imports 1985	Balance 1970	Balance 1985
Panama	0.3	0.5	14.6	54.8	–14.3	–54.3
Costa Rica	1.7	16.9	20.4	66.2	–18.7	–49.3
Nicaragua	3.8	5.1	2.6	9.9	1.2	–4.8
Honduras	13.1	32.2	13.9	24.9	–0.8	7.3
El Salvador	0.4	4.4	11.0	32.7	–10.6	–28.3
Guatemala	2.8	10.0	10.3	44.7	–7.5	–34.7
Total	22.1	69.1	72.8	233.2	–50.7	–164.1

Source: FAO 1977, 1988.

> Cutting the few valuable timber trees that can be found in a tropical forest on well-drained soils does not destroy the forest *per se*, although in the process when the trees are felled and dragged out of the forest, a large amount of forest destruction takes place. Usually about ten-times more trees of pole size or above are destroyed during the operation. (Budowski, 1990)

In many areas, though, soils are not well drained and in such conditions the use of machinery to extract timber can cause major environmental destruction. These effects have been noted by Baltissen in a study of logging operations in Costa Rica's northern Atlantic zone:

> The land degradation is mainly caused by the heavy machinery that is used; for instance, the bulldozer that enters the forest to remove the wood often cannot use the same road twice because of the high groundwater table and the bad carrying capacity of andeptic soils. So to remove only the valuable stems, 10 to 12 roads/ha are made by the bulldozer. (Baltissen, 1987:13, cited in Brüggemann and Salas, 1992)

Not all logging activities have such negative effects, however. This is particularly apparent in the case of artisanal forms of tree felling and pit sawing practised by farmers and Indian communities in many areas. Under these systems, trees are often sawn *in situ* and subsequently transported by cart with far fewer destructive consequences (ibid.).

Data for Costa Rica indicate that only 27 per cent of the wood which is cut actually reaches the market: 46 per cent is left in the forest to rot, while another 27 per cent is lost at the sawmill (Arcia *et al.*, 1991:46). This situation is partly explained by the use of inefficient technology and the fact that sawmill owners are reticent about investing in improved machinery and equipment when the future of the lumber business appears risky. It is also a result of the price structure and market situation for wood. Extraction and transport activities in Costa Rica account for approximately 75 per cent of the total value of the wood which arrives at the mill. The value of the standing trees represents just 8 per cent of the total. This price structure is conducive to highly selective felling of the biggest trees and most precious species.

Moreover, the fact that many tree species have little or no market outlets not only reduces the number of species which are exploited but also explains the scale of destruction and waste that characterizes felling operations (ibid.).

In Honduras, where the production of sawnwood averaged roughly 450,000 cubic metres during the 1980s, it has been estimated that 1.3 million cubic metres of wood were lost annually during the extractive process while approximately 300,000 cubic metres were lost during processing (Hernández and Ruíz, 1991). A large number of Honduran sawmills use outdated technology consisting of circular saws which process just 33 per cent of the wood which arrives at the plant (ibid.).

Nations and Komer make the point that 'the damage wrought by commercial logging is not so much the result of what foresters remove from the forests as what they leave behind – namely, the roads they construct to enter and exploit the area' (Nations and Komer, 1987:161). Along these roads follow peasant farmers and ranchers who rapidly convert forest to cropland and pasture, or land speculators who are likely to be closely allied with those engaged in logging operations. A speculator will mark out a specific area, provide advance money to peasants to clear the land and farm it for one or two years, and in the meantime deal with the national- and local-level bureaucracy to obtain legal rights to the land.[2]

THE EXPANSION OF EXPORT CROPS

As indicated above, deforestation in Central America is intimately tied up with the expansion of cash crops producing products destined for the export market. Between the early 1950s and the early 1970s, the area under export crops doubled from approximately 800 thousand to 1.7 million ha (Brockett, 1990; FAO, 1986). Three crops in particular – coffee, bananas and cotton – have had important implications for deforestation. The expansion of each was associated with the development of one of the three climatic regions described in Chapter 1 (see Map 2.1).

Coffee

Coffee production expanded primarily in central highland areas during the latter half of the 1800s in Costa Rica, El Salvador, Guatemala, and Nicaragua; and during the second half of this century in Honduras. In El Salvador, coffee came to dominate the entire economy. Following the growth of coffee production in the 1870s, El Salvador soon became the largest coffee producer in the region. The area under coffee increased from 57,000 ha to approximately 200,000 ha between 1921 and 1980 (Heckadon, 1989).

In certain countries – such as Costa Rica, El Salvador and Guatemala – the land acquired for coffee was already settled, if not densely populated (Brockett, 1990:23). In others, such as Nicaragua, the expansion of coffee brought large areas of sparsely populated land into production.

The fact that coffee plants were usually grown in combination with shade trees mitigated, to some extent, the negative environmental impact of coffee expansion. Moreover, coffee plantations constituted an important source of

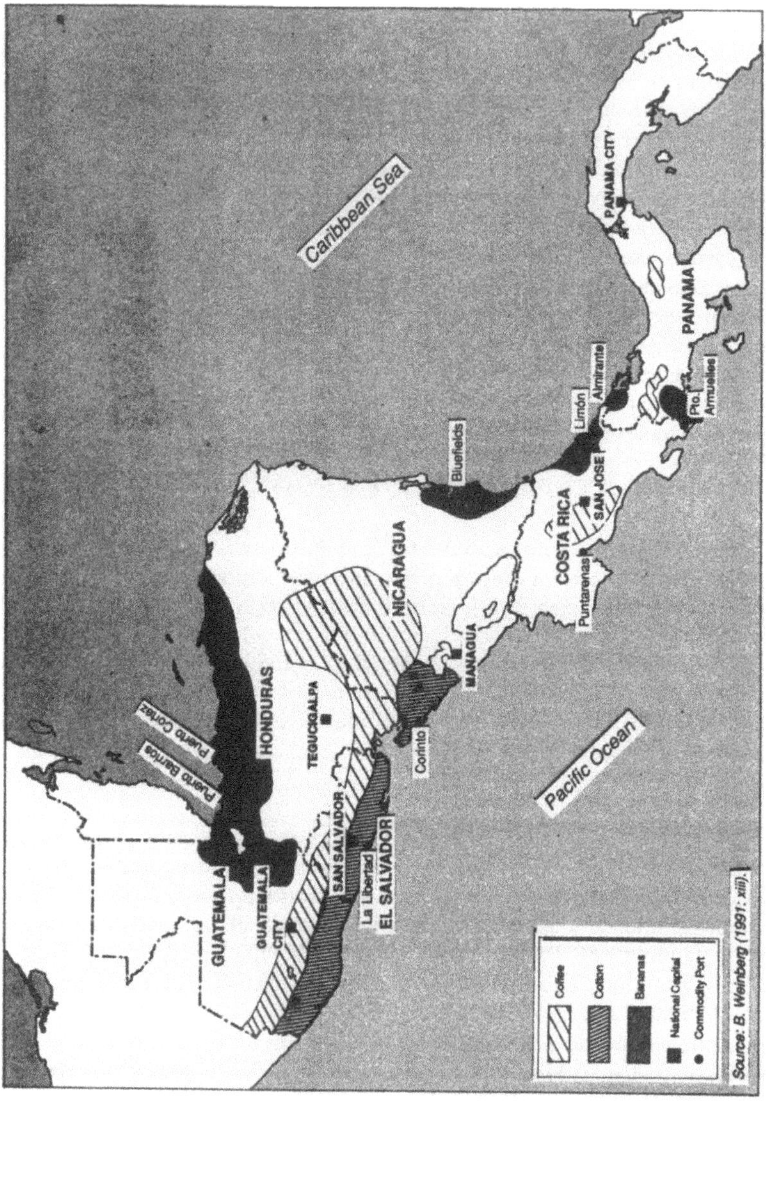

Source: B. Weinberg (1991: xiii).

Map 2.1 Central America: Agro-export zones

fuelwood.[3] The agroforestry system associated with coffee production went into decline in the late 1970s and early 1980s when government extension services and agribusiness corporations in a number of countries promoted the introduction of supposedly high-yielding coffee species which did not require shade. This not so 'green revolution' led to a sharp increase in deforestation and the use of agrochemicals in several coffee-producing areas. Faced, however, with falling prices on the world coffee market and indebtedness, certain producers have since reverted to the traditional system.

The linkages between coffee expansion and deforestation are most apparent in the following respects. First, coffee expansion was accompanied by considerable infrastructural development. As discussed below, the construction of road networks greatly facilitated the colonization of agrarian frontier regions.

Second, coffee expansion during the second half of the 19th century often occurred in areas designated as communal lands. In Guatemala

> Indians often lost their communal lands as land titling reforms were instituted to foster coffee growing and to promote and protect the European concept of private property. This notion was alien to the indigenous societies of Central America.... Although some Indians were able to protect their interests, many others were coerced or were cheated out of theirs in what ... has [been] called a massive assault upon village lands. (Brockett, 1990:23 citing McCreery, 1976:457)

Laws were passed – in Nicaragua in 1877 and in El Salvador during the early 1880s – which abolished, or encouraged the sale of communal lands (Brockett, 1990:25–6).

Thirdly, as access to communal lands was restricted and land concentration took place, large numbers of peasant producers were displaced and forced to clear more marginal lands, often located on forested hillsides, for grain production. Many had to migrate to towns or agrarian frontier regions. In Guatemala, migration was largely restricted up until the revolution of 1944 through a system of debt peonage and by highly repressive labour legislation which obliged Indians to work a certain period of the year for the coffee growers and other landowners (ibid.:23–4). In Costa Rica, the relatively high prices paid for prime coffee land in the Central Valley tempted many producers to sell up and buy large tracts of land in forest areas for ranching activities.

Bananas

Migration was sometimes in the direction of the banana plantations – which had been established during the second half of the 19th century in Costa Rica, and the early 1900s in Guatemala and Honduras. In Nicaragua, El Salvador and Panama, banana production remained relatively unimportant.

In the former group of countries, governments granted large concessions to (or what were to become) the United Fruit and Standard Fruit companies, mainly in sparsely populated Caribbean lowland areas (Woodward, 1985:177–83). The large scale deforestation which accompanied the expansion of the banana plantations resulted not only from the conversion of forest to crop land but also from the dramatic increase in demand for railway sleepers. Under the terms governing the concessions, the banana companies were required to take on the task of railroad construction.

By the end of 1918, the United Fruit company owned or leased half a million ha in the region. As Brockett explains, much of this land was virtually given away. Under an agreement which committed the forerunner of the United Fruit company to complete the railway linking San José with the Atlantic coast, that company was offered nearly a third of a million hectares of any state lands it chose as well as other generous concessions (Brockett, 1990:30).

The linkages between deforestation and banana expansion, however, involve far more than the simple fact that transnational enterprises cleared large areas of forest to make way for banana plantations and railways. The banana boom saw large numbers of labourers migrate to agrarian frontier areas, many of whom eventually acquired land in forest areas. This was partly due to unstable working conditions, the companies which employed them being unable to provide any guarantees for long-term employment. Plant disease and falling international prices have, periodically, led to recession or 'bust'. When laid off, many workers not only turned to forest clearance and subsistence agriculture as a means of eking out a living but would take up these activities with little knowledge of appropriate land-use practices in tropical forest areas.

Many others who continued to work in the banana plantations aspired to owning land and eventually becoming ranchers. In Costa Rica's northern Atlantic zone, for example, some would stake claims or buy land while continuing to work in the plantations in an attempt to obtain the capital necessary to take up farming (Brüggemann and Salas, 1992).

Cotton

Cotton production in the region soared from the early 1950s to the mid-1960s, experienced a six-year slump and then took off again until the 1980s. From the late 1950s, the number of cotton growers increased from 2,000 to 10,000 in less than two decades (Williams, 1986:31). Cotton production expanded in the Pacific coastal slopes and plains, on what are often the best soils in the region. Much of this land had already been cleared. In some cases, however, the expansion of cotton led to the rapid depletion of dry forest areas. The cotton boom of the 1950s and 1960s in El Salvador, for example, saw the virtual elimination of the last remaining areas of dry tropical forest in the Pacific plains (Heckadon, 1989). Between 1950 and 1965 the area under cotton increased from 19,000 to 122,000 ha.

The primary impact of the cotton boom on forest areas, however, was more indirect. As the area under cotton expanded, many peasants were displaced from the land while ranchers moved out of the Pacific area. As indicated earlier, these developments accelerated the colonization of agrarian frontier regions.

INFRASTRUCTURAL DEVELOPMENT

The expansion of the public road network greatly facilitated the colonization of forest areas. As indicated earlier, such infrastructural development had been intimately linked to the expansion of coffee production. Particularly important during the second half of this century was the construction of the Inter-American highway which was to link North and South America. Construction of this road accelerated during the early 1960s as part of the Central American Common Market initiative and by 1964 reached as far south as Panama (Woodward, 1985:272).

Costa Rica

In Costa Rica it has been estimated that the rate of deforestation on the Pacific side of the country increased five-fold following the construction of the Inter-American highway (Silliman, 1981:65). Similarly, deforestation on the Atlantic slope expanded rapidly following the building of the road linking the capital San José with the Caribbean coastal town of Limón (ibid.).

At the local level, farmers often enter into 'road-for-timber' agreements with loggers. How this process operated in Costa Rica is described by Budowski:

> In some areas, the various owners of forest land will happily sell out to logging companies because they will build a badly needed road, even if such a road may not last long since it was essentially built to take out only one crop of trees. In certain areas ... the local community which established itself (illegally) on the steep slopes made a contractual agreement with a local owner of a tractor. In exchange for opening the road he could take all the available timber growing relatively close to the road. Such an arrangement is extremely common in Costa Rica and many roads owe their origin to a 'road-for-timber' barter. (Budowski, 1990)

El Salvador

To understand the timing and scale of the deforestation process in El Salvador it is particularly important to refer to the process of infrastructural development in that country. As the historian Woodward explains, during the late 1800s and early this century 'El Salvador took the lead in road and bridge building. . . . [T]he network of roads gave the El Salvador of the 1920s the appearance of being the most progressive of the Central American states' (Woodward, 1985:161). Such developments clearly facilitated the ease with which migrants colonized forest areas and partly explain why El Salvador lost most of its forest resources several decades prior to other countries in the region.

Panama

Infrastructural developments both directly and indirectly account for much of the deforestation which has occurred in Panama. The attempt to complete the construction of the Inter-American highway, passing through the Darién region which links Central and South America, was an important factor conditioning contemporary patterns and rates of colonization in Panama (McKay, 1982; Sarmiento, 1985). Rapid colonization in western Panama during the early 1980s was facilitated by the opening of new access roads for mining and oil-pipeline maintenance (Jones, 1988:249).

Panama stands out from the other countries in that it does not conform to the traditional agro-export model. The crucial sector in the economy comprises service activities associated with the operation of the inter-oceanic canal. In 1980 agriculture only accounted for 10 per cent of GDP and the area sown with export crops was minimal. Nevertheless, throughout this century vast tracts of land have been cleared for pasture. The driving force behind this process was not so much the increase in the demand for beef in the United States but the expansion of the internal market which was linked to the intense process of urbanization and economic growth associated with the development of the Panama Canal Zone which dated from the early 1900s. As we will see below, such land-use practices led to serious environmental degradation and declining levels of agricultural productivity in many parts of the country. The potential threat posed by deforestation, however, is much greater. Land clearance, causing erosion and sedimentation of rivers and lakes, now threatens the very existence of the Panama Canal.

FUELWOOD AND URBANIZATION

The fuelwood trade has contributed to serious forest degradation in specific regions of a number of countries. A study published in 1984 (Dulin, 1984) found that the availability of fuelwood was 'very critical' or 'critical' in 38 per cent of the Central American region (excluding Nicaragua), notably in areas of high population density and in agricultural areas on the Pacific coastal slope (Martínez, 1986). The situation is particularly serious in the following countries or areas: El Salvador, where fuelwood accounts for 95 per cent of the consumption of wood products; around several of the region's major cities; the northern Pacific zone of Nicaragua; and in certain densely populated rural areas such as the western highlands of Guatemala.

The questions which have been debated internationally throughout much of the 1980s (Leach and Mearns, 1988; Foley, 1987) about whether or not the demand for fuelwood constitutes a fundamental cause of deforestation and whether there is in fact a fuelwood crisis (i.e. scarcity) are also pertinent to Central America.

One of the most comprehensive studies of the fuelwood situation of any one Central American country is that carried out by van Buren in Nicaragua (van Buren, 1990). The author argues that fuelwood is a waste or by-product of land conversion practices and that fuelwood collection should not be regarded as a major cause of deforestation. Abundance, rather than scarcity, characterized the fuelwood supply situation during the 1983/1984 period when the study was carried out. This was reflected in trends in relative prices, with the price of fuelwood rising less sharply than that of other products.

To what extent, however, was this a typical situation? With the benefit of hindsight, it is clear that the period of fuelwood abundance to which the author refers soon ended. It resulted from specific features of the post-revolutionary development model in Nicaragua which proved to be short-lived. These may be summed up in terms of state-centred accumulation, redistributive policies and urban bias (Utting, 1992). The ample supply of fuelwood to urban markets was partly accounted for by large scale land clearance projects undertaken by the state as part of a rapid modernization drive, supported by numerous external bilateral and multilateral development agencies. Moreover, these projects tended to be located in the more urbanized Pacific coastal region, thereby facilitating the flow of fuelwood to the principal cities where fuelwood consumption is high. The agrarian reform programme had confiscated *Somocista* properties belonging to the ex-dictator and others linked closely to his regime. Large state or co-operative farms were formed and considerable areas of land under tree cover were subsequently cleared to make way for crops and pasture.

These features of the Sandinista development model, however, were to change considerably during the latter half of the 1980s. Land clearance associated with large development projects slowed as the projects in question were either completed or delayed when funding dried up. The land redistribution process shifted away from state farms to peasant holdings located more in the interior of the country and towards the end of the 1980s a much tighter credit policy was introduced. It is likely that the amount of land clearance undertaken by private producers declined somewhat during the period of economic recession which characterized the latter half of the 1980s. Moreover, it is probable that demand for fuelwood increased significantly following the major increases in the cost of petroleum-based fuels (propane gas and kerosene) which followed in the wake of massive devaluations of the *córdoba* in 1988 and the elimination of subsidies.

There were clear indications by the end of the decade that deforestation of woodlands fairly near the capital Managua was accelerating as an increasing number of people turned to fuelwood gathering. This area, like much of the Pacific coastal region of Central America, experiences a long dry season of approximately six months when nothing much grows and when employment opportunities in agriculture decline dramatically. For many rural families, fuelwood collection provides an important source of income which enables them both to get through these lean months and to accumulate enough income to commence farming when the growing season begins with the onset of the rains in May or June. For many families, access to fuelwood has become increasingly difficult as local supplies dwindle.

There are indications that fuelwood gathering has contributed significantly to deforestation in other Central American countries. One estimate for Honduras indicates that annual collection and consumption of fuelwood accounts for an equivalent of 22,500 ha of forest (Campanella *et al.*, 1982:61). In El Salvador fuelwood gathering now constitutes the principal cause of deforestation and forest degradation (Heckadon, 1989). Up to three-quarters of all homes use fuelwood. Moreover, per capita fuelwood consumption is the highest in Central America (3.1 kgs per capita, per day), largely due to the low efficiency of traditional open-hearth cooking fires which have heat loss of up to 90 per cent (ibid.).

Studies conducted by CATIE indicate that nearly three-quarters of all Central American households (72 per cent) use fuelwood, with daily per capita consumption in the region at approximately 2.5 kilos (Martínez, 1986; Reiche, 1986). Of the region's total energy consumption, fuelwood accounts for more than half. Manufacturing, agro-industrial, food processing, and artisanal activities account for 31 per cent of total fuelwood consumption (Martínez *et al.*, 1983).

When considering the relationship between deforestation and fuelwood use, it is often important to distinguish between rural and urban fuelwood provisioning systems. In many rural areas fuelwood gathering is associated with more sustainable systems which involve the collection of small branches rather than tree felling. For this reason one should be wary of certain estimates concerning the correlation between fuelwood consumption and deforestation. Very often such estimates are calculated on the basis of the amount of fuelwood an average family consumes. Multiply this by a certain number of families and one arrives at a quantity of wood which corresponds to X number of trees or Y hectares of forest. If, however, branches are cut and the tree remains intact and relatively healthy, then obviously the relationship between fuelwood gathering and deforestation changes considerably. Where serious problems of deforestation do arise, it is often in connection with the fuelwood trade which operates to supply urban areas.

Rural to urban migration and high rates of population growth have led to a sharp increase in urban populations in recent decades. Urban growth in the region currently exceeds 4 per cent per annum (Merino and Mata, 1991). The situation is particularly alarming in Honduras and Nicaragua where rates of urban growth are of the order of approximately 5 per cent per annum.

Linkages between urbanization and deforestation in Costa Rica and the region as a whole have been identified by researchers from the Costa Rican Demographic Association. These relate primarily to forest clearance for urban settlement, increased demand for food, the displacement of agricultural producers to less productive lands, and the increased demand for wood for construction materials and fuelwood for industry, artisanry and household consumption (Arcia *et al.*, 1991; Merino and Mata, 1991).

While levels of per capita fuelwood consumption tend to decline in the context of urbanization and economic growth, fairly high levels of fuelwood consumption nevertheless persist in the more developed countries and urbanized areas of the region. In Costa Rica – the most prosperous country in the region – for example, half the population still uses fuelwood (Gewald, 1980; Martínez *et al.*, 1983). In urban areas generally, it is common for a third of the population to continue to use fuelwood. A Guatemalan study of fuelwood consumption in two categories of urban centres (above and below the level of 15,000 inhabitants) found that 32 per cent and 52 per cent of the families respectively continued to use fuelwood (Bogach, 1981).

Urban sprawl, the growth of commercial agriculture and the increased demand for fuelwood have led to serious deforestation in green-belt areas and hills which surround the region's major cities. The lives of millions of people living in or around urban areas have been affected by chronic shortages of water; health problems caused by the dust associated with wind erosion; the increasing cost of fuelwood; flooding of urban areas and the destruction of urban drainage and sewage systems; declining agricultural productivity in green belt areas; changes in micro-climate; and the increasing incidence of drought conditions, all of which can be linked to the process of deforestation around urban areas.

Some of the worst instances of erosion in Nicaragua are found in the rural areas surrounding the country's three largest cities, Managua, Leon and Chinandega. The fuelwood trade supplying Managua has led to serious deterioration of the capital's southern watershed. Throughout much of the past decade Managua experienced flooding as a result of silting of the city's drainage system as soil was washed down from the hills. Agriculture in the capital's green-belt area was also threatened by declining yields. In some areas in the surrounding hills erosion has been such that the soil cover is now too thin to grow crops.

The fuelwood trade has also affected other rural areas near Managua. The main source of the capital's fuelwood supply is found to the east of the city (Gewald, 1980; Reiche and van Buren, 1984). This is one of the most environmentally devastated areas in Nicaragua. Widespread deforestation over the last 20 years has not only affected agricultural productivity but led to changes in micro-climate. Drought conditions regularly ruin harvests. In recent years, this area has even been threatened by famine. In 1990, for example, 32 communities suffered chronic food shortages (*Barricada* newspaper, 21 October, 1990).

In October 1990, the Nicaraguan natural resources institute, IRENA, imposed strict controls on fuelwood extraction in the municipality of San Francisco Libre where extensive woodland areas have traditionally supplied a significant proportion of Managua's fuelwood requirements. IRENA, initially, declared a total ban on fuelwood gathering in the area. However, local authorities and residents protested, arguing that many families depended on the fuelwood trade for their survival. As a result, IRENA agreed to issue permits to local fuelwood cutters but banned operators from outside the municipality (ibid.). Controls were also imposed on trees and areas which could be cut while at the same time it was announced that a major reforestation project would be implemented in the area.

The situation is also dramatic in the northern Pacific region of Nicaragua where the cities of Leon and Chinandega are located. In this, the cotton heartland of the country, an important reforestation project was implemented during the early 1980s. In order to tackle the serious problems of wind and rain erosion which diminished soil fertility in the area, over 1,200 km of windbreaks were planted. By the end of the decade, however, many of the trees had been cut for fuelwood[4] (*Revista del Campo – Barricada* supplement, 12 January, 1990:4).

It has been estimated that the fuelwood sub-system in El Salvador involves collection and marketing circuits that extend up to 100 km from the capital.[5]

The fuelwood trade, centred on the capital San Salvador, threatens not only nearby parks like El Espino but also the mangrove forests along the Pacific coast. Destruction of the mangroves has damaged local marine life and the livelihood of coastal fishing communities. Pressure on fuelwood supplies has increased in recent years as a result of the acceleration of rural to urban migration during the war years. Moreover, it is common for migrants, including children, to cut and sell fuelwood as a means of acquiring money (UNICEF, 1988). To meet internal demand, imports of fuelwood (usually illegal) from Honduras and Guatemala have increased in recent years.

The cost of fuelwood has risen sharply during the past two decades. Many families must allocate an increasing proportion of their limited cash incomes to fuelwood purchases. Analysis of price data for Nicaragua indicates a significant change in relative prices of fuelwood *vis-à-vis* other basic consumer items and wages. In Table 2.4 it can be observed that whereas in 1982 the price of a bundle (*manojo*) of fuelwood was roughly half that of a pound of rice or beans, in early 1991 the gap had narrowed somewhat. Moreover, the number of bundles which could be purchased with the average wage fell from 17 to 8 over the same period.

The implicit labour cost of gathering fuelwood has also risen as families devote more time to gathering. In rural areas in Guatemala rural families were found to spend between half a day and two days per week collecting fuelwood (Zanotti, 1986). In certain areas of the northern Pacific region of Nicaragua and elsewhere in Central America fuelwood gatherers must now travel distances of more than 5 km to find fuelwood. Families living in communities which supply Nicaragua's second largest city León, reported that they began to collect and sell fuelwood during the late 1960s and early 1970s when land was cleared to make way for cotton. Ten to 15 years later,

Table 2.4 Prices of fuelwood, basic food products and the average wage in Nicaragua, 1982–91

Product	Measure	1982 (C$)[1]	1991 (C$)[2]
Fuelwood	Bundle	1.60	1.00
Rice	lb	3.15	1.20
Sugar	lb	1.70	1.00
Beans	lb	3.55	1.55
A Average wage		2,754	800
B Monthly family expenditure on fuelwood[3]		160	100
C A/B		17	8

Sources: 1982 data: van Buren, 1990.
1991 data: author's research.

1. Old *córdobas*.
2. New *córdobas*.
3. 100 bundles a month.

however, they were having to travel up to 12 km to find fuelwood in areas which were 20–30 km from the city of León (van Buren, 1990:88). By 1988, the Managua fuelwood trade circuits had already extended as far as the central interior department of Boaco some 100 km from the capital (ibid.:132).

This situation affects men, women and children alike. During the mid-1980s, it was estimated that in Nicaragua, 42 per cent of fuelwood gatherers were women or children (19 per cent and 23 per cent, respectively). In Guatemala and Honduras over a fifth of the time allocated to fuelwood gathering was taken up by women and children. Having to allocate increasing time to fuelwood gathering can have a negative social impact in terms of child care, the nutritional status of child gatherers, and education (Martínez, 1986). In certain rural areas of Costa Rica, where fuelwood collection is undertaken almost exclusively by men, this activity can account for 22 man-days each year (Reiche, 1986).

As shortages of fuelwood and the value of this commodity increase, so too do social tensions and conflict. In El Salvador it is common for those involved in fuelwood collection for commercial purposes to undertake these activities armed and at night. Under such conditions it becomes extremely dangerous for forest rangers or others to attempt to defend protected areas. Increasing social tensions have also arisen in rural areas between landowners and local fuelwood gatherers. Heckadon describes the situation in rural El Salvador, where traditionally fuelwood is regarded by the rural poor as common property:

> As shortages and prices of forest products increase, landowners become less tolerant of traditional practices. But even so, with or without the permission of landowners, the people continue to cut trees for fuelwood. In the coffee zones of the central and western regions, men, women and children mobilize at night . . . to cut shade trees for fuelwood. In rural zones, conflicts over the right to use the diminishing stock of forest resources constitute one of the principal sources of legal disputes. (author's translation, Heckadon, 1989)

The growth in demand for fuelwood and the increasing scarcity of fuelwood in certain regions, then, have had major ecological and social repercussions. The upshot of these trends has been: the serious degradation of significant areas of woodlands and mangroves, particularly those supplying urban areas; the failure of several environmental protection or rehabilitation schemes; increased costs, in terms of household expenditures and labour-time, for low-income families and, notably, women and children; as well as the growth of illegal activities and corresponding social tensions and conflicts.

Chapter 3

Structural and policy determinants of deforestation

AGRARIAN STRUCTURE AND LAND TENURE

Deforestation associated with shifting agriculture and the colonization of agrarian frontier regions is intimately tied up with processes of land concentration and dispossession which characterized the development of agro-export agriculture (Carriere, 1991; Goodman and Redclift, 1991). So-called 'liberal' states which emerged during the late 1800s and early 1900s, promoting private enterprise and restricted forms of representative democracy, fostered the growth of coffee production and introduced laws which tended to accelerate the disintegration of communal holdings. The profit opportunities associated with the cotton, beef and sugar booms of the 1950s, 1960s and 1970s, further intensified processes of land concentration and landlessness. By 1970, approximately half of all rural families were either landless or farmed sub-subsistence plots of less than a hectare (see Table 3.1).

It is important to note not only the extent of land concentration and landlessness which occurred following the 1950s but also the fact that the peasantry in most countries became increasingly marginalized during this period (Weeks, 1985: 117). In Honduras, for example, the percentage of farm

Table 3.1 Distribution of rural families by size of holding, 1970

Size of Holding (ha)	Guatemala (%)	El Salvador (%)	Honduras (%)	Nicaragua (%)	Costa Rica (%)	Central America (%)
Landless	26.6	26.1	31.4	33.8	26.3	28.1
Less than 0.7	15.0	24.4	10.3	1.5	32.2	16.8
0.7–4	42.3	36.2	24.1	24.2	13.1	32.6
4–7	6.9	6.2	11.9	7.9	4.8	7.4
7–35	7.4	4.9	18.1	18.1	14.6	10.7
35–350	1.4	2.0	3.9	13.5	8.3	4.0
More than 350	0.4	0.2	0.3	1.0	0.7	0.4
Total	100	100	100	100	100	100

Source: INTAL 1973; SIECA, 1981 (See Weeks, 1985, Table 34:112).

families with less than 1 ha increased from 10 per cent to 17 per cent between 1952 and 1974. In Guatemala the percentage of agricultural families with less than 1 ha doubled from 20.4 per cent in 1964 to 41.1 per cent in 1979. As Weeks explains:

> In the second half of the 1960s and the 1970s, Guatemala experienced a government-fostered process of marginalization of the peasantry to a degree perhaps unique in Latin America in such a short period of time. This process was integral to the further development of large-scale agriculture, concentrating land in large estates and increasing the size of the rural labor force to work those estates. (ibid.:117–18)

Among the six countries considered here, there are a number of important differences in agrarian structure and land tenure systems. Land concentration is particularly extreme in Guatemala and El Salvador. A study carried out in Guatemala in 1979 found that while 81 per cent of the registered 605,000 farm units were less than 3.5 ha, 60 per cent were less than 1.4 ha (see Barraclough and Scott, 1987:50). Pressure on the land is particularly intense in the north-west highland area of the country. At the beginning of the 1980s, this region accounted for 40 per cent of the national farm population but just 16 per cent of all farmland (ibid.). A similar situation prevailed in El Salvador until the 1980s. In 1980, just 3 per cent of all farm families controlled 80 per cent of the land. At the other end of the scale, landless labourers made up 60 per cent of all farm families while another 35 per cent occupied 'micro farms' which were too small to meet the subsistence needs of the household (ibid.: 68).

Agrarian structure in Panama, Costa Rica, Nicaragua and Honduras was somewhat different given the existence of a significant group of middle-sized commercial farmers. In Nicaragua, for example, 32 per cent of all rural families worked holdings of between 7 and 350 ha (see Table 3.1).

This latter feature of rural social structure has important implications for our analysis in that deforestation, associated in particular with the expansion of pasture land, was linked to processes of social differentiation among the

Table 3.2 Land concentration in the 1970s

Country	% of holdings[1]	% of farmland[1]	% of rural families with 4 ha or less[2]
Guatemala	2.6	65.5	83.9
El Salvador	1.5	49.5	86.7
Honduras	4.2	56.0	65.8
Nicaragua	1.8	46.8	59.5
Costa Rica	9.1	67.2	72.6
Panama[3]	6.7	66.1	58.4[4]

Source: 1. Census Data reported in Brockett, 1990:72.
2. See Weeks, 1985, Table 34 A, p. 112.
3. FAO, 1986:43.
4. 2.9 ha or less.

peasantry. The conversion of forest to agricultural land occurred, to some extent, via a process of 'extensive accumulation' involving the acquisition of sizeable tracts of land in agrarian frontier regions. As a result of this process there emerged a significant sector of 'rich peasant' producers.[1]

The existence of this sector had important cultural implications in that it reinforced popular perceptions which saw the forest as a resource to be cut down and converted to other land uses. Here was living proof that the process of displacement from the land, migration and colonization of frontier regions did not necessarily reproduce poverty but, for some at least, held up the possibility of social advancement. Moreover, the fact that many of these producers were ranchers, reinforced the notion that the key to success lay in cattle raising.

While land concentration was also acute in Costa Rica and Panama, there were significant differences in land tenure systems. This is particularly apparent in the case of Costa Rica. Unlike the four Central American countries to the north, the development of commercial agriculture in Costa Rica did not occur in the context of a coercive system of labour relations which tied peasant and landless labourers to large landowners and excluded the mass of the rural population from political participation (ibid.:119). As Weeks explains, the dispossession of smallholders from the land which accompanied the expansion of commercial agriculture occurred largely by economic means. Higher land prices induced many peasants to sell their lands. As labourers, many could earn a living wage due to the fact that scarcity of labour pushed up wage rates (ibid.).

Considerable changes in land tenure took place in Panama during the 1970s. At the beginning of that decade, the state purchased approximately a quarter of the country's agricultural land and gave land titles to 10,000 new peasant holdings and 11,000 'squatters', a total of 23 per cent of all farms (Barraclough and Scott, 1987; FAO, 1986).

Throughout the region much of the inhabited land area remains untitled. Because of the itinerant character of agriculture in many frontier regions, the low value of land, the complicated legal and administrative procedures for obtaining titles, and the costs involved, few bothered about obtaining legal titles (Augelli, 1987:5; CEDARENA, 1990).

The use and possession of untitled land have important implications for deforestation and forest protection schemes, several of which are identified in a study of the Tortuguero region of north-eastern Costa Rica (ibid.). First, to prove possession land is usually cleared of trees. Land clearance, in fact, serves two purposes. It acts as a precondition for obtaining legal title and as a defensive mechanism to prevent encroachment by land-seeking peasants or graziers who would be less likely to stake claims on land which shows signs of being settled (Kapp, 1989). Second, producers using untitled land are ineligible for bank loans. This both restricts patterns of land use and encourages producers to engage in activities requiring limited operational capital such as selling logs and raising cattle. Third, without security of tenure there is little incentive to engage in long-term natural resource protection practices. Moreover, in Costa Rica forestry bonds are only issued for reforestation projects on titled lands. Finally, government regulations to regulate land and resource use are difficult to implement in situations where land surveys and legal possession are non-existent.

GOVERNMENT POLICY AND LEGISLATION

Numerous laws and government policies, often backed up or shaped by the policy approach and funding priorities of international development agencies, have played an important role in deforestation processes (Nations and Komer, 1987; Healy, 1988). So too has government inaction in certain areas. In the case of Central America, this is particularly apparent in the following contexts.

Lumber exploitation

Throughout this century concessions have been granted for commercial lumber exploitation with little or no regard for reforestation and sustainable forest management. As explained by Budowski when examining the causes of deforestation in Costa Rica:

> ... there is no management, no silvicultural treatments to favour natural regeneration or the growth of existing regeneration such as valuable saplings or of other trees that have not yet reached commercial size. Indeed the forest is typically visualized as a non-renewable resource and logging of valuable species is often labelled as a 'mining' operation. (Budowski, 1990)

The environmental impact of such an approach was particularly dramatic in pre-revolutionary Nicaragua where the operations of United States lumber companies left large areas of the northern Atlantic coast region deforested. The situation in Honduras, where many forestry sub-system activities are controlled by the state, appears to be equally dramatic. According to a USAID report: 'Logging is marked by waste and inefficiency at every step from the stump to the mill. In the broadleaf forests only about 10 per cent of a given stand is logged' (Campanella *et al.*, 1982:10). The report goes on to highlight the 'lack of operational standards and procedures governing what slopes and soils can be logged without incurring site degradation, or where and how logging roads should be constructed' (ibid.:11).

A similar situation is described by Silliman when referring to the case of Costa Rica at the end of the 1970s:

> Although Costa Rica's forests are being felled at a rapid rate, very little of the wood is put to use. Estimates of commercial use of wood range from 10 to 40 per cent of that cut. The rest is burned or left to rot where it falls. (Silliman, 1981:67)

However, not only economic factors but also the contradictory effects of forest protection legislation account for this situation:

> The reasons for this high rate of wastage appear to be both economic and political. The cost of transporting timber from remote areas where small farmers cut it to sawmills and markets is high. Many native trees have no known commercial value, or at least no present market in Costa Rica. In addition, the

> Forest Law of 1969, designed to control forest exploitation, may have inadvertently encouraged waste. By requiring a permit to sell wood, but not to burn it or let it rot, it encourages waste by small farmers unwilling to obtain a permit. And, by prohibiting the export of unsawn trunks, it has discouraged a potential export trade which could increase the value of timber to the woodcutter. (ibid.)

Despite mounting criticism within Costa Rica of traditional patterns of forest exploitation, there was little change in the practices of the lumber companies during the 1980s. The 1990 Forestry Action Plan criticizes the degree of wastage in the lumber industry claiming that between 20 per cent and 30 per cent of what is cut is actually marketed (MIRENEM, 1990:8).

In Honduras, regulations governing tree felling have often remained on paper. According to one government report, the system governing the marking and marketing of trees has been affected by corruption (SECPLAN *et al.*, 1990:143).

Agricultural support services

Governments in the Central American region have generally organized agricultural support services for the benefit of commercial farmers. Credit policies, in particular, have been instrumental in encouraging the rapid conversion of forest to pasture land.

As indicated earlier, in the case of Panama the clearing of land for pasture has constituted the major cause of deforestation in that country. Underpinning these patterns of land use has been the government's credit policy. As Heckadon explains:

> Another factor influencing the *campesino* emphasis on cattle ranching is the manner in which credit for production operates in Panama. Development institutions providing loans to *campesinos* such as the Banco Nacional de Desarrollo (BDA) and the Banco Nacional de Panamá (BNP) generally favour ranching over agriculture. Private banks also prefer livestock development programmes because they require less supervision, less paperwork and lower risks while providing greater short-range financial returns than agricultural projects. (Heckadon, 1985:220)

The Central American cattle 'boom' of the 1960s and 1970s was underpinned by the massive injection of foreign aid and investment, notably from the World Bank and the Inter-American Development Bank. According to one estimate, the value of cattle sector projects supported by these agencies amounted to approximately $500 million by the late 1970s (Howard, 1987a). Once beef imports to the United States began to increase during the early 1960s, the government of that country actively encouraged bilateral and multilateral aid agencies, as well as private financiers and investors, to channel resources towards the cattle sector (ibid.).

The prioritization of commercial farmers and corporations in government agricultural support programmes has had the effect of restricting access by the mass of rural producers to essential goods and services for agricultural production, thereby fuelling the forces underlying migration and colonization of agrarian frontier regions.

Colonization projects

Government policies have, until recently, openly encouraged the colonization of rainforest areas both as part of a modernization strategy and also an attempt to ease social pressures. Colonization has been primarily of the 'spontaneous' kind although a number of large planned colonization schemes have been implemented. These include, for example, the Rigoberto Cabezas project in the southern interior zone of Nueva Guinea in Nicaragua; the Río Guayape project in the eastern central part of Honduras which was intended to cover 78,000 ha; and the Northern Transversal Strip in Guatemala where initial projects covered 50,000 ha although plans were formulated which, if implemented, would have encouraged colonization in an area of more than 880,000 ha (Jones, 1988:245).

The Rigoberto Cabezas project was drawn up by the Somoza government partly in response to the escalation of agrarian conflicts in Pacific coastal areas. During the 1960s international banks such as the Inter-American Development Bank (IDB) began to provide finance for colonization schemes. This policy had been promoted by the United States government as part of the Alliance for Progress strategy which, in the aftermath of the Cuban revolution, encouraged several governments throughout Latin America to implement reforms as a means of defusing the threat of revolution. The government also saw the area as a potential bread basket and major producer of cattle for export. The feasibility studies barely even acknowledged the 'ecology question'. Yield estimates for products like beans were highly optimistic – approximately two to three times higher than those achieved in practice during the 1980s (IAN, 1972; MIDA, 1980).

In countries like Costa Rica and Panama government policies actively supported so-called spontaneous colonization (Jones, 1988:245). As Augelli observes when writing about Costa Rica:

> Until the 1970s the same perception of land abundance, held by private citizens, influenced government planning and policies. The goals of the government were to populate the national territory, to increase farm output, and to defuse the problems of land hunger in the densely settled areas where commercial crops were produced. In the past the government frequently used public lands to pay municipal debts and to support construction of a transportation network (Augelli, 1987:4).

Even during the early 1980s, the law required colonists to clear the forest in order to acquire right of possession. Following the Spanish legal code, legal right to the land was established on the basis of 'peaceful and continuous possession' (ibid.). Taxes on naturally forested land were higher than those on 'improved' land (Silliman, 1981:65).

To deal with the increasing incidence of land invasions and illegal squatting which had intensified during the 1950s, the government passed, in 1961, the Law of Land and Colonization which barred such actions and set up an institute to deal with land redistribution on an organized basis. By 1983 some 22,000 families had received nearly 800,000 ha of land. As Augelli points out, however: 'In keeping with the traditional perception of land abundance, the legislation assumed that much of the acreage for redistribution would be virgin forest areas still in the public domain' (Augelli, 1987:16).

While several Costa Rican laws sought to facilitate small farmers' access to land, it was often the case that large landholders ended up being the principal beneficiaries (ibid.:14). The 1942 Law on Squatters' Rights, for example, intended, as Augelli explains:

> to compensate landholders in the Central Valley whose lands had been invaded and occupied by squatters. Under the terms of this law, the government compensated such owners with parcels of equal value elsewhere in the country. Because frontier lands were much cheaper than those in the Central Valley, many large landholders used this law to exchange small plots in the Central Valley for immense tracts on the frontier. Some large holders even paid people to become squatters on their property in order to demand compensation from the government. (ibid.)

A similar situation occurred in Guatemala where the Northern Transversal Strip was officially designated for colonization by landless Indians (Brockett, 1990:71). During the 1970s, however, wealthy landowners and military officers acquired large tracts of much of the best land in the area, which some referred to as 'the zone of the generals' (Riding, 1979 cited in Brockett, 1990:71).

Legislation governing land titling in Costa Rica also encouraged deforestation. A provision of the Law of Ownership Information (Ley de Informaciones Posesorias) allows claimants to title up to 300 ha if land is to be used for cattle raising and only 100 if the land is to be used for agriculture. According to a report by CEDARENA:

> Reforestation or management of natural forest are not mentioned in this article of the law.... [T]he lure of tripling one's land in a single *información posesoria* doubtless encourages the wider destruction of forest and planting of pasture. (CEDARENA, 1990)

The same law stated that people wishing to obtain title to areas where the majority of the land was covered in forest must provide evidence in the form of a public document issued more than 10 years earlier. This also had negative repercussions: 'As requiring a 10-year-old public document made lying about possession of forested area much more difficult, people simply cut 50 per cent or more of the forest cover' (ibid.). This loophole was subsequently pasted over by the 1986 Forestry Law but a new one emerged:

> Possession for the purposes of titling property is now much easier to demonstrate than before and also easily falsified.... Once proof of ... possession has been arranged and the land titled, logging permits – for those who bother to seek them – may be easily obtained. Inadequate supervision of logging means that deforestation will often follow. (ibid.)

DEFORESTATION IN THE 1980s AND 1990s: WAR, AGRARIAN REFORM, RECESSION AND ADJUSTMENT

A number of changes in the forces underpinning deforestation in Central America have occurred during the 1980s and early 1990s. Four phenomena, in particular, have had important implications for deforestation, namely: war, agrarian reform, the world recession and economic stabilization and adjustment programmes.

War

Throughout the 1980s, Guatemala, El Salvador and Nicaragua experienced civil wars. In certain instances, war has restricted the rate of deforestation as war zones are usually located in agrarian frontier or forest areas. Given the risks to human life and economic infrastructure – as well as the disruption of marketing circuits – peasants, commercial farmers and logging companies generally refrain from penetrating such areas. Moreover, hundreds of thousands of people fled the war zones or were obliged by the state to resettle elsewhere (as in Nicaragua and Guatemala). Such processes have relieved population pressure and stemmed processes of land concentration in agrarian frontier areas. Figures on displaced persons in El Salvador range from half to one million people. In Guatemala the number of refugees from civil violence was put at 220,000 in 1989 (Bradley *et al.*, 1990b). In Nicaragua an estimated 350,000 people migrated internally or fled into neighbouring Costa Rica and Honduras (Barry and Serra, 1989).

Displacement, however, can merely exacerbate problems of deforestation in neighbouring countries. Over 40,000 Guatemalan refugees are located in neighbouring Mexico. The number of El Salvadorean and Nicaraguan refugees believed to be living in neighbouring Honduras at any one time oscillated between 30,000 and 50,000 during the 1980s (SECPLAN *et al.*, 1990:54). During the early 1980s between 18 and 30 thousand Miskito Indians fled Nicaragua for Honduras. While attempts were made to locate much of this population in camps, many dispersed to take up shifting cultivation and cattle raising in the forested areas of eastern Honduras (Campanella *et al.*, 1982). By the late 1980s it was estimated that farming and military activities, as well as fuelwood gathering and human settlement patterns of some 10,000 Nicaraguans, had caused serious environmental destruction in an area of 400 square km (SECPLAN *et al.*, 1990:55).

Military operations have themselves caused major degradation of forest areas. Both the El Salvadorean and Guatemalan armies have operated a scorched earth policy. While information regarding the situation in Guatemala is hard to come by, it is known that military operations have caused widespread deforestation in certain highland regions such as Quiché and Huehuetenango. Extensive strips of forest lining roads in rural areas have also been cleared by the army to diminish the risk of ambush by guerrillas.

In El Salvador guerrilla strongholds in forested areas around the Guazapa Volcano, Morazán, Cabañas, and Chalatenango have, according to one report, 'become virtual wastelands, with the landscape scarred with bomb craters, crops destroyed, and burnt forests reduced to secondary scrub or rock' (Hall and Faber, 1989:9). Community leaders from a town in Morazán placed an advertisement in a national newspaper which read:

> We are very worried by the damages caused by the devastating forest fires caused by aerial bombing and indiscriminate mortar fire, as well as by soldiers carrying out patrols and operations. The armed forces commonly burn the forest during the dry season, accelerating the destruction of resources in the zone.... Because they have deforested large areas in our zone, the scorched earth and bombing campaigns have notably affected rainfall patterns. The length of the rainy season has shortened, and the levels of streams and rivers has dropped. The situation is becoming more critical, and we are worried now because it is affecting us directly. Our crops have diminished and this worsens our already agonizing economic situation. (*El Mundo*, 24 February 1988, cited in Hall and Faber, 1989:9)

War has also prevented governments and private agencies in the region from implementing reforestation, forest protection or agroforestry projects. This is partly due quite simply to the difficulties and risks involved in working in frontier or forest areas affected by fighting. It is also a result of the fact that governments having to contend with war and economic crisis are unlikely to prioritize environmental programmes.

Nicaragua not only experienced a civil war but also a trade and aid embargo imposed by the United States in 1985. One effect of this was that important environmental studies and projects supported by USAID in most countries in the region were not carried out in Nicaragua.

The ending of civil wars can also have major implications for deforestation. As the director of the Nicaraguan Natural Resources Institute (IRENA) remarked: 'It is a curious fact that the war was kind to the forest while peace is destroying it'.[2] He was referring to the fact that the rate of deforestation had declined as a result of the war but had increased sharply once the war ended in 1990. The demobilized *Contra* rebels and thousands of 'returnees' were reluctant to resettle in rural development poles established by the new government and were gradually moving back to agrarian frontier areas. Also, United States and Taiwanese companies had requested large territorial concessions to extract lumber, and a number of new national companies had been set up to export hardwoods to the United States.

Agrarian reform

Agrarian reform programmes, along with economic recession and military conflict, have introduced a number of important variations in agrarian structures and land tenure systems in the past decade. It is extremely difficult, however, to generalize about the implications of these developments for deforestation. The precise effects would need to be studied in specific local contexts.

Certain changes have occurred which would appear to augur well for deforestation in that they deflate, to some extent, pressure on the land associated with agrarian frontier expansion and migratory agriculture. We noted above that agrarian systems in Central America were characterized by high levels of land concentration and insecurity of tenure. During the 1980s, land concentration declined significantly in El Salvador and Nicaragua, and partially in Costa Rica and Honduras (Baumeister, 1991). Moreover, there has been a marked increase in the number of owner-occupiers and a corresponding reduction in the relative proportion of peasants and farmers who rent and sharecrop land or occupy national or communal lands (ibid.:21).

Improved access to land in countries such as El Salvador and Nicaragua, however, has often accelerated processes of land clearance. In the case of Nicaragua, for example, it might be assumed that improved access to land, security of tenure, the availability of cheap credit and modern farming inputs would have alleviated the pressures on shifting cultivators and hence acted as a brake on the process of deforestation on hillsides and in frontier regions. In certain areas of the country, however, the process of land redistribution took little account of the type of land which was given to peasant producers. As such, it appears that significant areas of forest land were handed over to peasant producers and subsequently converted to crop or pasture land. As the director of IRENA remarked in an interview: 'Too much emphasis was placed on the idea of agrarian reform while no one paid attention to the need for an agroforestry reform'.[3]

The Salvadorean agrarian reform, which (according to official 1988 estimates) redistributed over a third of a million hectares to nearly 80,000 beneficiaries between 1980 and 1987, proved to be even more ecologically unsound. An estimated 45 per cent of the land available for redistribution under Phase I (which nationalized several large estates) was so-called 'non-productive' land with some woodland or forest cover (Hall and Faber, 1989:8). Much of this was cleared by co-operatives for agricultural purposes. Less than 30,000 ha were officially retained as reserves.[4]

By dividing their estates among family members many large coffee plantation owners were able to block implementation of Phase II of the reform programme. This meant that much of the country's prime agricultural land was unavailable for redistribution. Instead, a land-to-the-tiller programme (Phase III) gave land titles to peasant farmers who had previously rented or squatted on marginal plots. Hall and Faber make the point that this had the effect of 'locking them into the same plots for 30 years instead of traditional shifting cultivation [thereby] eliminat[ing] the crop rotation and fallow periods essential for soil recovery' (Hall and Faber, 1989:8). Lack of support services for the programme's beneficiaries and increasing indebtedness have forced many to overexploit the lands they received (ibid.).

World recession

The world recession has had important implications for deforestation. During most of the 1980s, the Central American economies generally stagnated while several countries, notably Nicaragua, recorded negative growth during several years. Stagnant or falling international prices for sugar, cotton and beef, and restrictions on Central American imports into the United States put a brake on the expansion of agro-export agriculture. A positive development from the ecological point of view was the fact that the area under pasture did not expand during the 1980s. Value added in agriculture actually declined between 1979 and 1985 and grew by just 5 per cent over the decade (Baumeister, 1991). Traditional agricultural exports fell from $3,016 million in 1979 to $2,165 million in 1989 (ibid.).

In Chapter 2 we noted the way in which processes of capital accumulation and marginalization associated with the agro-export 'model' underpinned deforestation during the second half of this century. While the agro-export recession no doubt served to constrain the 'accumulation dynamic' which resulted in vast tracts of forest land being cleared to make way for agro-export production, it may well have intensified the 'survival dynamic' and obliged thousands of families to intensify the exploitation of forest resources. Particularly important in this context is the fact that recession in agriculture has put a brake on the process of proletarianization of the rural labour force (ibid.). In several countries, there has been a marked decrease in the level of demand for harvest labour in the agro-export sector.

Economic stabilization and adjustment policies

Economic stabilization and adjustment policies have also had important implications for deforestation. Partly as a result of these policies, so-called 'non-traditional' agro-export sectors (flowers, spices, vegetables, sesame, cocoa, etc.) have expanded. The ecological question marks associated with this type of production have more to do with the intensive use of pesticides than with increased land clearance and deforestation. In general, such production involves the more intensive use of land than that which characterized traditional agro-export production (ibid.). Moreover, due to the perishable nature of non-traditional commodities and the need for multiple inputs, processing and storage facilities, production generally occurs in areas having relatively sophisticated infrastructure and which are close to urban centres. This usually implies distance from forest areas. In some areas, however, the expansion of non-traditional crops has displaced basic grain producers further towards forested hilltops and the agrarian frontier. This is particularly apparent in certain highland regions of Guatemala.

The fuelwood crisis has been compounded by economic stabilization policies attempting to eliminate distortions in relative prices deriving from subsidies and exchange rates which over-value local currencies. To the extent that currency devaluations and the reduction or elimination of subsidies increase the cost of petroleum-based fuels this is likely to increase the demand for fuelwood.

Stabilization programmes intent on cutting fiscal deficits have also imposed restrictions on government expenditures. In some cases this has severely limited the budgets of environmental agencies. In 1988, the Nicaraguan Natural Resources Institute (IRENA) not only experienced significant budget and personnel cuts but also lost its ministerial status. Becoming yet another department of the Ministry of Agriculture, it lost considerable influence in decision-making processes. In 1991, the Honduran government announced a 15 per cent reduction in budget of the state forestry corporation

(COHDEFOR). While governments throughout the region have come under increasing international pressure to implement forest and environmental protection projects, state agencies often lack the necessary resources to fulfil even basic counterpart obligations, let alone launch and effectively implement major conservation initiatives.

PART II
THE BREAKDOWN OF TRADITIONAL RESOURCE MANAGEMENT SYSTEMS

Overview

The preceding analysis identified a series of processes and activities, associated with particular patterns of accumulation and subsistence production, which resulted in what can be described as an assault on nature. The picture which emerges is one of the unchecked clearance of forest land and the use of inappropriate land-use practices and technologies – in short, of society's blatant disregard for nature and any notion of sustainable development.

This is only part of the story, however. Historically, throughout the Central American isthmus, certain Indian groups and *ladino* peasant farmers practised a variety of farming or resource management systems which were more appropriate in ecological terms. But many such systems have collapsed. Others are currently breaking down, confronted with the effects of market forces, government policies, land concentration, population growth, migration patterns and economic crisis.

The following three chapters analyse the breakdown of traditional resource management systems and examine how the livelihood of people located in or close to deforested areas is affected when such systems collapse. The analysis focuses on three specific resource management systems associated with two social groups: Indian populations and *ladino* peasants.

Regarding the former, two very distinct resource management systems are examined. One concerns the more stereotypical system generally associated with Indian groups in sparsely populated rainforest areas, where swidden agriculture is combined with hunting and gathering. The analysis, based on secondary sources, refers primarily to the situation of groups in Costa Rica and Nicaragua.

A second indigenous resource management system, analysed in more depth, is that associated with Indian groups in the department of Totonicapán in Guatemala's densely populated western highland or *Altiplano* region. This is an area where families own extremely small plots of land but where, historically, forests have been protected as a result of a relatively effective system of communal regulations governing the use of forest resources. An in-depth study of this area, undertaken by Veblen during the early 1970s, identified the various social and cultural relations and practices underpinning forest protection. Since the publication of Veblen's thesis (see Veblen, 1978), the case of Totonicapán has often been sited in the literature on deforestation and forest protection as a 'model-example' of how local communities can manage natural resources on a sustainable basis and as a 'counter-example' of both the tragedy of the commons thesis[1] and neo-Malthusian theories which draw a strong link between population growth and environmental degradation.

The situation described by Veblen has, however, changed somewhat during the past two decades. Communal forest protection mechanisms are now considerably weaker and deforestation has increased. A case study of this area was undertaken to identify the factors which were threatening this resource management system and to examine the impact of these changes on the lives of the local population.

The third resource management system examined below is that associated with forms of shifting agriculture based on crop–fallow rotation as practised by certain groups of *ladino* peasant producers. This analysis, largely based on secondary sources, refers primarily to the case of peasant producers in Panama and Nicaragua.

Chapter 4

Deforestation and Indian populations

CENTRAL AMERICA'S INDIAN POPULATION

Estimates of the size of Central America's Indian population vary widely, from approximately 3.5 million to 5 million. Most of this population lives in Guatemala where approximately half the country's inhabitants are Indian. In other countries of the region the Indian population generally represents less than 10 per cent of the national population. Table 4.1 presents one set of high-range estimates taken from several reputable sources.

In several countries, much of the Indian population is located in or near forest and agrarian frontier regions. As is well known, indigenous households and communities often depend on the forests for numerous products. In the case of the Kuna Indians of Panama, for example, it has been reported that:

Table 4.1 Indian population of Central America (late 1980s)

Country	Indian population	% of national population
Panama	170,000[1]	7.5
Costa Rica	24,000[2]	1.0
Nicaragua	80,000[3]	2.0
Honduras	208,000[4]	3.5
El Salvador	500,000[5]	10.0
Guatemala	4,000,000[6]	47.0
Total	4,982,000[7]	16.6

Sources:
1. Burger, 1987.
2. Bozzoli, 1986. This figure includes groups living in or around Indian communities with a considerable degree of 'ladinization'.
3. Solis, 1989.
4. SECPLAN *et al.*, 1990, reports a total Indian population of 208,095.
5. There is little reliable statistical information on the size of the indigenous population of El Salvador. Some estimates put the figure below 100,000 (see IICA/FLACSO, 1991, Table 7). An ethnographic study conducted in 1975 (Marroquín, 1975) estimated that 10 per cent of the population could be classified as Indian. Applied to today's population figures, this would mean an indigenous population of approximately half a million (see Chapin, 1989).
6. Chapin, 1989.
7. Another 30,000 Indians, representing 16 per cent of the national population, live in Belize (Burger, 1987).

> Hunting and gathering of products from the forest play major roles in their subsistence lifestyles. A . . . study of just one small part of the reserve identified seventy-two agroforestry combinations, utilizing forty-eight trees and sixteen crop plants. The Kuna use thirty-six species to build boats, thirty-two for fuelwood, forty species for home construction, and scores more for medicine, handicrafts, and utensils. (Barborak and Green, 1985:6)

A study of several indigenous reserves in south-eastern Costa Rica describes the diversity of plants used by the Indian population:

> For fuelwood, Guacimo (*Guazuma ulmifolia Ham.*), Pochote (*Bombax ellpticum*) or Palo de Mora (*Chlorophora tintoria*) are used. Among the numerous medicinal plants can be found Madero Negro (*Giniicida maculata HBK*) used for skin irritations and Tuete (*Veronica patens HBK*) used as a cure for parasites. Multiple purpose tree species also exist such as Cimarron (*Spondias spp.*) which provides fruit, bark to heal wounds and wood for construction. The wood of the balsa tree (*Ochroma logpus SW*) is used to manufacture rice sacks, the branches are painted in health rituals and the flowers are dried to provide filling for pillows. (Wunderlich and Salas, 1991:19 – author's translation from Spanish)

The penetration of lumber companies, plantation and mining enterprises, ranchers and peasant farmers into agrarian frontier areas has often had a dramatic impact on the livelihood of indigenous groups. Many Indian groups live in conditions of extreme poverty as a result of the loss of customary rights governing land use and the exploitation of forest resources; the exploitative character of labour relations in extractive or farming enterprises; the marginalization of indigenous populations in most government social programmes; and vulnerability to the difficult physical conditions which characterize many rainforest areas.

While Costa Rica enjoys one of the highest levels of social development in Latin America, with relatively low rates of infant mortality and high rates of literacy, much of the Indian population lives in a state of extreme poverty. Official estimates indicate that approximately half of those over seven years of age remain illiterate (Bozzoli, 1986), although other reports put the figure much higher (*Gente*, 1990:6). An FAO report on rural development in Panama estimates that some 70 per cent of the 80,000 Guaymí Indians live in a state of 'critical' poverty (FAO, 1986:7).

Referring to the situation of indigenous populations on Nicaragua's Atlantic coast before the 1979 revolution, Jenkins writes of extreme rates of illiteracy and levels of infant mortality, frequent epidemics of measles, typhoid, chronic malaria and widespread venereal disease: 'In synthesis, it can be said that the health conditions of the indigenous population of the Atlantic were the worst in the country' (Jenkins, 1986:128). Many of the social gains (notably in health care and literacy) achieved during the early 1980s were reversed in the latter half of the decade as war, economic crisis and cuts in government expenditure took their toll.

A government report on the environmental situation in Honduras noted that:

> the living conditions of the [indigenous] population are far worse than those experienced by the rest of the rural population. . . . The long history of discrimination, exploitation, and marginalization of ethnic groups has resulted in progressive [cultural, social and economic] disintegration. (SECPLAN *et al.*, 1990: 72 and 74)

Writing about the Indians of El Salvador, Chapin observes: 'They are invariably poor, illiterate, and without opportunities to better their standard of living' (Chapin, 1989:30–31). Moreover, their situation has worsened during the past decade:

> Before 1980 the Indians constituted the Salvadorean underclass – the 'poorest of the poor', in the language of development agencies – and they were in no position to take advantage of the benefits of the agrarian reform. Not only have they been excluded from the reform; it is clear that they have had their traditional access to small plots of land for subsistence farming as well as seasonal labor on the estates almost entirely blocked. In short, their poverty has worsened significantly over the last decade. (ibid., 62)

Referring to Guatemala in the mid-1980s, Burger writes:

> Guatemala is one of the few countries in the world where the indigenous population is in a majority. However, their standard of living, conditions of work and present political persecution are unparalleled anywhere. (Burger, 1987:76)

STATE POLICY AND INDIAN RIGHTS

Government policies and legislation regarding the rights and protection of Indians vary considerably throughout the region. At one extreme is the case of Guatemala where, as one USAID report notes:

> . . . from Independence to 1985 [when civilian government was formally restored], the Mayan lifestyle was regarded by the dominant culture as an obstacle to the social and economic development of Guatemala. The state maintained an integrated cultural policy with the objective of dissolving the Mayan culture. (Bradley *et al.*, 1990b, Vol.II, 20)

The dissolution of indigenous cultures and traditional resource management practices has also resulted from sometimes well-intentioned but short-sighted government development policies which fail to comprehend the mechanics of indigenous society. A relevant example is that provided by Houseal *et al.*, referring to the Embera Indians of the Darién rainforest region of Panama. With support from the United States Department of Agriculture, the Panamanian government established, in 1966, an inspection zone along the country's southern border with Colombia to prevent the spread of foot and mouth disease and thereby safeguard Central Amercian beef exports to the United States. As explained by these authors:

> One of [the] first actions . . . was to prohibit free-roaming, hoofed animals in the twenty-five mile inspection zone along the international border. Hence, the Embera's pigs, their main export commodity and source of hard currency, were

> abruptly eliminated with no viable alternative provided to these people. The effect of the policy has been to induce the Embera to become lumbermen, a lucrative occupation encouraged by the national government, but one which would jeopardize the remaining Darién forest and the survival of Embera culture. (Houseal *et al.*, 1985:14)

A similar lack of understanding of indigenous society characterized Sandinista development policy towards Nicaragua's Atlantic coast region, home to several ethnic groups. Following the Sandinista revolution which toppled the Somoza dictatorship, the new government set about tackling the historic problem of the isolation and marginalization of the Atlantic coast by emphasizing the need for rapid economic modernization. What emerged was a technocratic approach to deal with a highly complex 'ethnic question'. Investment policy took little account of local forms of production and trade, and was characterized by the design and implementation of large scale development projects, often incompatible with the socio-economic and demographic characteristics of the region (Vilas, 1989). The disruption of traditional lifestyles and production systems intensified in 1982 when approximately 40,000 people were forced to abandon their communities as the government established free-fire zones to combat armed insurgents.

In all countries, laws intended to protect the rights of Indians remain inadequately enforced and ineffective. Even in Costa Rica where progressive legislation defends Indian rights to the land, the forces of modernization and the market tend to overwhelm those of protection. After many decades, during which Indians were dispossessed of their lands, the Costa Rican government introduced progressive legislation during the 1970s to defend Indian land rights. The 1977 Ley Indígena set aside five areas totalling 100,000 ha as indigenous reserves and prohibited access by outsiders (Burger, 1987:69). By 1990, 21 reserves existed. Belonging to ten different ethnic groups, they are located almost entirely in remote areas in the south of the country.

Implementation of the law, however, has left much to be desired and encroachment has proceeded apace. Referring to the situation in the mid-1980s Burger observes:

> Since 1977 commercial logging, land invasions by non-indigenous settlers and various mining operations have occurred on the reserves. Oil exploration is currently taking place on Talamanca land, a hydroelectric project is planned on the Borunca reserve and with the growth of production of sugar cane and coffee, communal land is being claimed as within the public domain and being planted with cash crops. In 1983 the Asociación Indígena de Costa Rica claimed that a North American company had obtained a permit to develop 10,000 hectares of indigenous forest. (Burger, 1987:69)

Of the nearly 3,000 ha which form part of the Guatuso reserve in the north, only 400 ha belong to the Indians who reside there, while just 8 per cent remains under forest (Wunderlich and Salas, 1991:16).

The Costa Rican state has done little to intervene in disputes between

Indians and outsiders settling on reserve land. While the law defends the rights of Indian communities, it fails to deal with the territorial conflicts which threaten the lands, livelihood and lifestyle of Indian communities (Current K., 1991:43). The judicial process to resolve disputes regarding the illegal occupation of Indian land is extremely slow while the National Commission for Indigenous Affairs (CONAI), with responsibility for recovering lands, has been granted few funds and personnel (Wunderlich and Salas, 1991:34).

ENCROACHMENT IN TROPICAL RAINFOREST AREAS

Pre-Columbian Central America was home to numerous Indian groups who engaged in sustainable forms of subsistence provisioning. Pressures on the natural resource base were minimized through a combination of conditions related to low levels of surplus extraction (due primarily to the absence of markets), low person to land ratios, as well as cultural perceptions and religious beliefs which led people to respect nature's elements.

Agricultural practices based on long crop–fallow rotations ensured the regeneration of forest areas cleared for cultivation and the recuperation of soil fertility. Migratory agricultural practices were generally combined with hunting, gathering and fishing. Since the arrival of the Spaniards in the 16th century, these systems have been under threat (Fournier, 1991), with many Indian groups forced to migrate from coastal or lowland areas to more inaccessible highland regions. Despite the effects of colonization and modernization, several indigenous groups still practise sustainable resource management systems. As Houseal *et al.* explain:

> ... native peoples have devised sustainable long-term land-use practices combining migratory agricultural practices with arboriculture and wildlife management. Each group has developed a distinct subsistence economy that provides their foodstuffs, construction materials, textiles, medicines, utensils and other physical needs. In addition, they have marketed a variety of agricultural, livestock, wildlife and fisheries products to the outside society for decades. Their mixed agricultural and forestry systems produce more labor, more commodity per unit of land, and are more ecologically sound and result in more income distribution than any other practices currently imposed upon their lands. (Houseal *et al.*, 1985:10)

The stresses affecting such systems have intensified dramatically during the latter half of this century due primarily to encroachment by outside forces. When analysing the impact of encroachment on indigenous groups, two scenarios stand out: that relating to the impact resulting from settlement by ranchers and *ladino* or white peasant farmers, and that resulting from the penetration of large 'extractive' enterprises (notably lumber and mining).

INDIANS, GRAZIERS AND COLONIZATION IN COSTA RICA

As pointed out by Current, referring to the situation of rural land invasion or 'squatting' (*precarismo*) in Costa Rica:

> *Precarismo* ... is certainly destroying Costa Rica's best protectors of forest: the Indians.... The whites who occupy Indian land often exploit the Indians by setting fire to their fields and houses, poison their water supply, let animals loose to trample crops or get them drunk until they sign a bill of sale. (Current, K, 1991)

Overt coercion, however, is only one of several mechanisms which has had the effect of separating Indians from their means of production and led to the breakdown of traditional resource management systems. Land was often acquired through less oppressive means by taking advantage, for example, of the Indian's lack of knowledge regarding land prices and/or the belief that land was abundant and could be acquired elsewhere (Wunderlich and Salas, 1991:9). In certain reserve areas, local officials of the Costa Rican Land and Colonization Institute (ITCO), established in the 1960s to administer the Indigenous Reserves and regularize the colonization process, also exerted pressures on Indian families to sell their lands.

Very often Indians were displaced to more marginal and inaccessible areas characterized by higher rainfall, steeper slopes and poorer soils (Houseal *et al.*, 1985:10). Not only are many families finding it increasingly difficult to practise sustainable resource management as land becomes more scarce but indigenous communities are rapidly losing the knowledge base associated with such practices. This has resulted partly from absorption into alternative economic activities, such as wage labour, but also from processes of acculturation ('ladinization') and out-migration which affect in particular the younger generations.

By 1990, 40 per cent of the 322,000 ha offically classified as indigenous reserves in Costa Rica, were occupied by non-Indians. Roughly the same area (44 per cent) had been deforested, largely as a result of land clearance for cattle grazing by non-Indians (see Table 4.2).

This process of encroachment has forced much of the Indian population to abandon traditional agricultural or subsistence provisioning practices based on long crop–fallow rotations and hunting and gathering. A study of six reserves in south-eastern Costa Rica revealed that many Indian families now possessed small farms of less than 5 ha, and often *minifundia* of less than 1 ha, while it was common for many non-indigenous farmers to possess holdings of between 100 and 200 ha (Wunderlich and Salas, 1991:16).

Graziers in these reserves often leave areas of their ranches unfenced, with the result that cattle graze in adjacent lands belonging to Indians and in the

Table 4.2 Costa Rica: Land tenure and forested areas in indigenous reserves, 1990

Reserve area (ha)	Indian land (%)	Non-Indian land (%)	Forested area (ha)	Forest area/ reserve area (%)
322,103	60	40	181,005	56

Source: CONAI (See Wunderlich and Salas, 1991, Table 1, p. 16).

process destroy crops. With the expansion of cattle grazing there has been a marked increase in the number of forest fires. This has resulted largely from graziers setting fire to pasture before the onset of the rainy season in order to encourage new grass and to 'clean' pastures of weeds, snakes, ticks and so forth.

The same study indicates that many landless Indians or *minifundistas* work as labourers or watchmen for non-indigenous farmers, often receiving less than half the official minimum wage. Intra-community conflicts have also intensified as a result of certain Indian families occupying lands held by other Indians or selling lands claimed by others (ibid.:17). Access to fuelwood and medicinal plants has become restricted, and sources of subterranean and surface water have diminished significantly (ibid.:22). Impoverishment and the loss of cultural identity have resulted in increased alcoholism, drug addiction, squatting and even practices such as grave robbing (ibid.:18).

INDIANS AND LOGGING COMPANIES IN NICARAGUA

Given the failure of most lumber companies and national governments to undertake significant reforestation and adopt sustainable forest management practices, the operations of logging companies in any particular area often assume a 'boom and bust' character. This was particularly apparent in the Atlantic coast region of Nicaragua where United States companies cleared an estimated 240,000 ha during the early and mid-1900s. As Solis points out:

> While in 80 years of activity US logging companies extracted more than one billion board feet from the Atlantic Coast, the value of exports between 1945 and 1964 alone were worth an estimated 50 million dollars...., However, because the Somoza administration required no investments in restocking or reafforestation, in environmental protection, nor in infrastructure and productive activity that might provide long-term employment to break the vicissitudes of the boom-and-bust cycle, the extractive nature of the Atlantic Coast enterprises had a long-term impact on the local population. (Solis, 1989:491)

Lumber companies would move in, clear large areas and then move out. Such was the case, for example, of the Nicaraguan Long Leaf Pine Lumber Co which began operations in 1945 and soon became the largest lumber company in the country. By the 1960s it had already depleted much of Northern Zelaya's pine resources and it ceased operating in 1963 (CIERA, 1981:50).

The arrival of the lumber and other resource extraction companies transformed the livelihood systems of local indigenous populations who previously had formed part of subsistence economies dedicated to hunting, gathering, fishing and domestic agriculture. The Indian population was quickly drawn into the market economy both as wage labourers and as consumers. In 1925, the Bragman's Bluff Lumber Co was Nicaragua's largest employer with nearly 3,000 workers (ibid.:49). By the 1950s, lumber companies in the Atlantic coast region employed tens of thousands of workers.

The cultural and socio-economic implications of this change were dramatic. As hunters, gatherers, fishermen and subsistence agriculturalists, work in the traditional Miskito communities had been irregular, itinerant and often combined with leisure. This was a far cry from the conditions which characterized work in lumber activities (ibid.:67).

Many Indians chose to work for the lumber companies on a seasonal basis, returning to their communities until economic hardship obliged them to work again as wage labourers. The need to return was to a large extent related to the fact that Miskito families had been drawn into the market economy. As manufactured products, wages and new cultural influences penetrated Northern Zelaya, purchases of clothing, tools, metal cooking utensils and tinned foods increased, often replacing traditional products.

Many would return to find work as wage labourers given the fact that the capacity of certain families and communities to generate cash incomes on the basis of certain traditional activities decreased. Historically, trade had been limited to the exchange of skins, medicinal plants and tortoise shells for items such as clothing and arms. The devastation of large forest areas and the growth of coastal populations which accompanied the lumber and banana booms, reduced the availability of many traditional marketable products.

Miskito Indians would also return to their communities once the foreign companies pulled out. Partial 'proletarianization' and the seasonal character of employment were important factors preventing the breakup of indigenous culture. Men would generally return to their communities on a seasonal basis or when company operations ceased in a particular area. Women, who had stayed on in the communities, played an important role in maintaining traditional lifestyles and cultural values and transmitting these to children (Jenkins, 1986:126).

When analysing the situation of Indian populations and their relations with outside forces there is a tendency to generalize about the plight of the Indian. A number of anthropological studies, however, reveal the extent of social differentiation within Indian tribes or communities and the way in which certain groups may ally with outside agents for economic and political gain. This has important implications for the analysis of the social impact of deforestation. A case study of the Xicaque Indians of Honduras (the findings of which are presented in Part III) found, for example, that several of the chiefs or *caciques* who controlled the various tribal groups appropriated the bulk of the benefits which derived from the exploitation of forest resources and certain development programmes and projects.

Encroachment has also resulted in considerable social conflict. In Nicaragua, for example, an estimated 100,000 people moved into agrarian frontier regions between the 1950s and the 1970s. This process not only led to environmental deterioration but also to conflicts over land. As Solis observes:

> ... the spiralling depletion of natural resources was accelerated as the expanding indigenous population was forced to move along the rivers also seeking new lands. The inevitable friction resulted in some deaths prior to [the Sandinista Revolution of] 1979, specifically of Indians killed by Spanish-speaking migrants ... (Solis, 1989:492)

Encroachment, then, has had a dramatic impact on the livelihood systems of local indigenous populations. It has reduced the areas available for shifting agriculture based on long crop–fallow rotations, displaced Indian families to more ecologically fragile and/or isolated areas, fundamentally transformed social relations of production and indigenous culture, and generated social tensions and conflict.

Chapter 5

Deforestation and livelihood in Guatemala's western highlands

Certain rural areas of Central America, notably in Guatemala and El Salvador, are characterized by high population density and extreme land scarcity. Many such areas were deforested long ago. In some parts of the western highlands of Guatemala, however, local communities were able to protect their forest resources and considerable areas of dense pine or mixed forest still remain. In recent years, however, these forests have been subjected to intense pressures and traditional resource management or forest protection systems show signs of breaking down. Since the vast majority of the local population depends heavily on the use of forest products this situation has had a significant impact on people's livelihoods. To analyse the breakdown of these systems and the social impact of deforestation in such areas, a case study was undertaken in the department of Totonicapan and the neighbouring municipality of Nahualá (department of Sololá) in the *Altiplano* region of Guatemala (see Map 5.1).[1]

Guatemala's largest area of dense pine forest is located in the department of Totonicapan. Until fairly recently the rate of deforestation in the area remained relatively low. This situation was particularly unusual given a number of socio-economic and demographic characteristics of the area.

First, with a population density of 260 inhabitants per square km, Totonicapan is one of the most densely populated departments of the country.

Second, there is considerable demand for forest products, not just for fuelwood which is used by virtually the entire population, but also other products. Totonicapan produces the bulk of the country's inexpensive 'down-market' pine furniture. Several thousand artisans process large quantities of white pine (*Pinus ayacahuite*).

Third, *minifundismo* in the area is rife. While less than 2 per cent of all farms account for nearly 30 per cent of agricultural land, land concentration in the area is far less extreme than in most other areas of the country. In Table 5.1, we see that just 2.8 per cent of the country's farmers control 66 per cent of agricultural land. Moreover, large landholdings of over 100 ha are extremely rare in Totonicapan.

The extent of land scarcity in the area is reflected in the fact that the standard Central American unit of land measure – the *manzana* (0.7 ha) – is rarely used. People refer instead to the number of *cuerdas* they own – a *cuerda*, in this particular part of the *Altiplano*, being one-sixteenth of a *manzana*, or 437 square metres (approximately 21 × 21 metres).

Map 5.1 Guatemala: Location of Totonicapán and Sololá

The last agricultural census of 1979 indicates that 69 per cent of all holdings were less than 1 *manzana* (see Table 5.1). Through time, small family holdings have been constantly subdivided, resulting in a situation whereby it is common for families in many areas to possess just two to five *cuerdas*. In certain parishes of the municipality of Nahualá the average holding is nine *cuerdas* (Leal, 1989). Land in more densely populated rural areas such as the Valley of Totonicapan is now extremely difficult to purchase. Land prices have risen from 300–400 Quetzales (approximately $100) a *cuerda* in 1985 to 1,500 Q ($380) in 1990. In terms of a daily wage equivalent, the price of land has increased from 75 to more than 200 days work.

Despite pressures such as these the forests of Totonicapan were not subjected to the type of intense processes of deforestation characteristic of

Table 5.1 Agrarian structure in Guatemala and Totonicapan, 1979

	Guatemala				**Totonicapan**			
Size[1]	**No of Holdings**	**%**	**Area (mz)**	**%**	**No of Holdings**	**%**	**Area (mz)**	**%**
Total	468243	100	5562468	100	27451	100	40492	100
1c–1mz	154140	32.9	71389	1.3	18856	69	7794	19
1–2mz	100540	21.5	135923	2.4	4401	16	6246	15
2–5mz	107950	23.1	322365	5.8	2744	10	8325	21
5–10mz	46074	9.8	305404	5.5	945	3	6366	16
10–32mz	37814	8.0	669831	12.0	439	1.6	7061	17
32–64mz	8655	1.9	384011	6.9	50	0.2	2256	6
1–10cab	11750	2.5	1767649	31.8	15	–	1506	4
10–20cab	861	0.2	748573	13.4	1	–	938	2
20–50cab	369	0.1	682387	12.3	–		–	
50–100cab	73	–	315818	5.7	–		–	
100–200cab	15	–	126662	2.3	–		–	
200+cab	2	–	32458	0.6	–		–	

Source: Agricultural Census, 1979
1. c = *cuerda* = 437 square metres; 16 c = 1 mz
mz = *manzana* = 0.7 ha.
cab = *caballería* = 64 mz = 45 ha

Guatemala and other areas of the Altiplano throughout this century. This unusual state of affairs was analysed by Veblen during the early 1970s. He attached considerable importance to the strength of communal structures, inter-personal 'face-to-face' relations, and the fact that local residents who depended heavily on the use of forest products, notably carpenters and artisans, organized to ensure that the use of forest land was regulated[2] (Veblen, 1978). Also important was the fact that pressure on forest resources was relieved somewhat by the existence of diverse employment opportunities. It is common for families in Totonicapan and Nahualá to engage in a combination of agricultural, artisanal and commercial activities. While most families have small plots of land, many also allocate time to making furniture, weaving traditional Indian fabrics or making western-style clothes by way of a home work system. Official employment statistics for Totonicapan indicate that 36 per cent of the economically active population work in agriculture, 34 per cent in the artisanry and 20 per cent in commerce (SEGEPLAN, 1991).

This particular mode of production relieved pressure on forest resources not simply because a variety of employment opportunities existed, but because the economic system ensured that value added along the chain of extraction, production, processing and marketing remained, to a large extent, within the local economy and was not siphoned off by external agents. This situation applied in particular to the timber sub-system. In certain communities, for example, it was common for families to obtain permission to fell two trees a year. One tree would be used to cover the family's need for fuelwood or timber. The other would be used as payment to a local woodcutter. The woodcutter was likely to produce planks to sell to a local carpenter or to supply the family carpentry business. The local carpenter would produce furniture which would be marketed by a local merchant or another family member.

Compared to many other areas of the *Altiplano*, this economic system proved fairly effective from the point of view of basic subsistence provisioning. Moreover, a significant number of artisans were able to accumulate sufficient capital to invest in other activities, notably commerce and their children's education (Smith, 1990).

Since the mid-1970s, however, it is apparent that the situation, described by Veblen, relating to low rates of deforestation, has changed significantly. Veblen reports that the main area of dense forest cover (along the María Tecún Ridge) decreased only 7 per cent between 1954 and 1972. A 1972 satellite image of the area shows dense forest cover spreading along four spurs running west to east. All four connect to a central ridge on their western flank, also under dense forest cover. Very little cleared space is to be seen between each spur and no significant differences in the extent of forest degradation are visible along each of the spurs. In a 1988 satellite image each of the spurs is more clearly defined as a result of deforestation in the areas which separate them. Also the two southern spurs are much paler, indicating considerable degradation of the forest area.

Statistical interpretation of the 1988 images has not yet been conducted for this area but a number of other studies confirm the above observation. A study conducted by the Faculty of Agricultural Sciences of the University of San Carlos apparently found that the area under forest cover had declined by some 14 per cent between 1974 and 1979.[3] Two micro-level studies of small forest areas in the municipalities of Totonicapán and Nahualá indicate a loss of forest cover of 15 per cent and 40 per cent between 1972 and the late 1980s, respectively (Arreaga, 1990; Granados, 1989).

The Totonicapán case study sought to examine what had changed in the past two decades to prompt this reversal of trends, and also to assess the impact of deforestation on the livelihood and lifestyles of a population that depended heavily on the use of forest products. The study found that this change in the use of forest land had resulted from, first, the weakening of traditional forest protection mechanisms and secondly, the increasing exploitation of forest resources. The latter had occurred as a result of the rise in demand for fuelwood and timber from outside the department, along with a fairly generalized crisis of petty commodity production in the area which had forced many to turn to the exploitation of forest products as a means of obtaining supplementary income.

THE LAND TENURE SYSTEM AND THE DEMISE OF COMMUNAL PROTECTION MECHANISMS

There are four main types of forest property in Totonicapán (see Map 5.2). These are communal, municipal and private forest and what are known as the

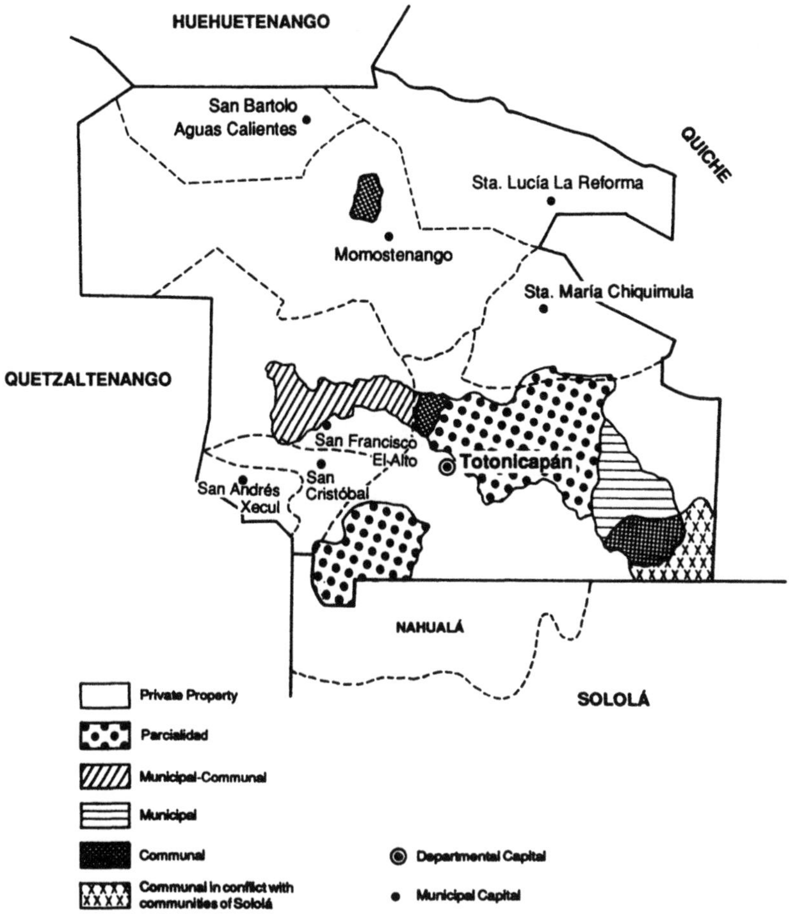

Map 5.2 Totonicapán: Forest property regimes

parcialidades where land belongs to extended kinship groups or clans. It is important to analyse deforestation in the context of these different property forms for each can be associated with contrasting attitudes and practices governing the exploitation of forest resources and forest protection.

The communal forests and the parcialidades

Much of the dense forest area of Totonicapán is communal forest or belongs to the *parcialidades*. Throughout this century, communal protection mechanisms in these areas have played an important role in preserving the forests.

Various types of protective mechanisms appear to have prevailed. First, measures were taken to keep out outsiders and check on who was using forest products (Veblen, 1978). Each community elected forest guards or rangers who had considerable authority and respect. In forest areas under threat from the activities of outsiders, the community in general would take an active interest in protecting the forest and informing the forest rangers of any suspicious activity.

Regular community assemblies were organized to discuss problems and possible solutions regarding deforestation and other community concerns. These meetings also acted as a check on the performance of local leaders who could be removed if accusations of corruption or malpractice were accepted by the community at large.

Such meetings, however, are generally attended only by men. While this situation raises obvious questions regarding women's rights and needs, some women consulted in the course of this study did not feel particularly excluded. Rather than feeling subordinated in the local power structure, they interpreted men's attendance at such meetings in terms of the husband's role in the domestic division of labour. As family spokesperson, he was expected to convey the views of his household following a discussion with the wife and other family members prior to the meeting. In the community of Paqui, for example, we asked one woman whether she felt excluded from the community decision-making process. She replied: 'Of course not. I tell him what to say and it is his duty to go to the meeting to report our views. If men didn't do this, what use would they be.'

Communal forest protection mechanisms were particularly strict in the *parcialidades.* Only members of the clan had the right to use the forest and anyone wanting to fell a tree had to obtain permission from the leaders or *principales.* Those who did not comply with the regulation were punished. Such punishments might include, for example, suspension of the right to graze sheep or collect leaf litter for fertilizer in communal forest areas; several days work in reforestation activities; or the imposition of a fine which might contribute to the maintenance costs of the tree nursery.

Second, as noted above, in many communities the tradition existed whereby each family could fell two trees a year to provide for household fuelwood requirements and construction materials. Often one of the trees would be used as payment to the person who felled the tree and chopped or sawed the wood.

Third, in certain communities, sustainable management practices were adopted whereby the best reproductive specimens were left standing. Selective, as opposed to clear felling was generally practised. But even in areas that had been cleared for agriculture it was sometimes the case that a large healthy specimen would be left standing to facilitate regeneration once soil fertility declined and agricultural production was abandoned. In the course of this study researchers came across certain areas under forest cover which were once used for agriculture. On the ground could be seen the faint outlines of terraces which perhaps a half century earlier had been cultivated. In the middle of the terraced area would stand a large tree, possibly a hundred years old.

These practices not only facilitated natural regeneration but also ensured the persistence of mixed forests which were more resistant to disease and permitted the perpetuation of more valuable species such as white pine. It was factors such as these which enabled the pine forests of Totonicapán to resist the onslaught of pine borer disease which affected significant areas of pine forest in the western highlands in the late 1970s and was often used as a pretext by the state forestry corporation and lumber companies to clear-fell large areas.[4]

Cultural perceptions of the Mayan-Quiché Indians also played an important role in the system of forest protection. According to traditional beliefs, both human beings and 'natural' elements, such as trees, possess a spirit. The existence of one complements that of the other and each has a predetermined role in the universe. The place and function of the forests in the universal order are as important as those of humankind. Nature must be respected and what is taken from it should be replaced. In the past, for example, it was common for families in many communities to plant several seedlings for every tree they felled. Traditional Mayan-Quiché culture saw humankind as one actor among many and the different elements which made up the universe were to be respected and preserved (Castañeda, 1991). The interdependence of the natural elements and the respect accorded to trees in Mayan culture is indicated in the following extract from an ancient Mayan document:

> Our lord made the heaven and the forest. In the heaven he sowed the stars and in the forests he sowed trees. The roots of the stars and the trees are one and the same. When a tree falls, so too does a star . . . (Chan K'in Viejo de Nají, cited in Diechtl, 1988 – author's translation from Spanish)

During the past two decades, several communal protection mechanisms have ceased to function or have become less effective. This has been due to a variety of factors: the emergence of a new state regulatory code; both endogenous and exogenous pressures on forest resources; and the ongoing demise of traditional Mayan cultural beliefs.

The practice of assigning two trees per family appears to have died out. During the past two decades this communal protection mechanism disappeared as state authorities attempted to introduce a licensing system to regulate tree felling. In theory, families who want to fell trees submit a request to the community authorities who study the different petitions, accept, reject or modify them accordingly, and then request one or several licences from the state forestry corporation, DIGEBOS. Once the licence is obtained, the trees are allocated to the respective families. In practice, however, many communities fail to comply with the licensing system, partly for reasons associated with the cost of licenses but also due to the fact that obtaining them can be time-consuming. The heavy-handed attitude of INAFOR (the state forestry corporation which existed between 1967 and 1982) in implementing forest regulations, also served to alienate many communities. While the role and attitude of the forestry corporation changed significantly once INAFOR was disbanded and a new entity – DIGEBOS – was created, many communities are still suspicious of the state agency with responsibility for regulating the use of forest land.

Apart from institutional problems such as these, there is increasing pressure on the use of forest resources both by local residents and outsiders. As will be analysed in more detail below, what was once a bustling local economy of peasants, traders and artisans has been hit by recession. Unemployment and underemployment have increased and many of the self-employed have had to abandon their family enterprises and look for work locally as labourers. Others have migrated in search of work in other departments, Mexico or the United States. Subsistence provisioning of food has become increasingly difficult as family plots are sub-divided, the availability of organic fertilizer decreases and the cost of agrochemicals increases. Moreover, Totonicapán, unlike certain other departments of the Altiplano has not been on the receiving end of large amounts of development aid, often associated with the expansion of non-traditional cash crops such as vegetables.

In this context, it is to be expected that local residents look increasingly to the forest and its 'free' raw materials as a source of income or to reduce costs. While many community leaders are genuinely concerned about the problem of deforestation, they do not believe that the local residents are primarily responsible. A recurrent theme to emerge from meetings and talks with community leaders and local residents during the course of this study, was that many local people have no other alternative but to exploit the forest for subsistence purposes. Those activities which should be controlled, they believed, involve the illegal exploitation of forest products for relatively large-scale commercial gain.

Such activities are associated more with the lumber industry than with fuelwood production. In recent years, however, a new menace has emerged – illegal bark stripping. The way in which this activity is organized has outraged the inhabitants of several communities, particularly the *parcialidades*. The bark of the white pine has become a valuable commodity. It is used as a softening agent for curing leather in the tanneries of Huehuetenango to the north of Totonicapán. Until the latter half of the 1980s, these enterprises generally used imported materials (*quebracho*). However, recurrent devaluations of the *quetzal* (Q), associated with economic stabilization and adjustment programmes, caused a dramatic increase in the price of imported *quebracho* from 2,400 Q/ton some five years ago to 9,500 Q/ton in early 1991. Smaller tanneries in Huehuetenango were particularly affected by this price rise. Looking for substitutes, they turned their attention to the pine forests of Totonicapán which were more accessible by road than those of the northern *Altiplano*. Moreover, traditional forest protection and management mechanisms practised by the Indians of Totonicapán had ensured the persistence of mixed pine forests and a significant presence of white pine. As mentioned earlier, such practices were particularly effective in the *parcialidades*. It has been in these areas that the bark strippers have concentrated their activities.

These activities assume a clandestine and criminal character. Bark stripping takes place at night, usually when there is a full moon to provide natural illumination for night-time work. The men involved are often armed and there have been cases where local residents or forest guards have been tied up, threatened or actually injured.

During the full moon period, two to three trucks, laden with pine bark, reportedly leave Totonicapán each night for Huehuetenango. Contracted locally by truck owners, gangs of men move into an area and in two to three hours can strip dozens of trees. The men usually work in teams of two. In two to three hours a team can amass four *quintals* (400 lbs) of bark from approximately 20 to 24 trees. Each man is paid between 60 and 80 Q ($12–16) per *quintal* and in one night can earn at least 120 Q ($ 24), the equivalent of approximately 15 full days work. The bark is sold in Huehuetenango for around 300 Q ($60) a *quintal* or approximately 6,600 Q a ton. This represents a considerable saving on the price of imported materials.

There have been a number of instances where up to 500 trees have been stripped in one night. A count taken in six communities in early 1991 revealed that 5,580 trees had been stripped. The people involved originally concentrated their activities in more accessible areas along roadsides. During late 1990, however, this pattern changed and now the bark strippers tend to penetrate deeper into the forest in search of larger trees.

In a meeting with the governor of Totonicapán, promoted by officials from the state forestry agency and attended by UNRISD researchers, approximately 50 community leaders voiced their concerns about bark stripping in a series of eloquent interventions which vividly expressed the fear of local residents for both their physical safety and livelihood. These leaders were taking advantage of a recent change in government to approach the newly appointed governor, expose the situation and denounce the injustice and corruption of state officials who tolerated such activities.

They explained that there had been no effort on the part of the state authorities to curtail these activities. The police rarely interfered and in any encounters with the bark strippers, it was usual for those transporting the bark to bribe their way out of difficulty. When fines were imposed, they were extremely low. As one community leader explained:

> The law favours those who are doing the damage rather than those who are the victims. No one would mind paying a fine of 200 Q (per truck load) when they are carrying 20 to 30 *quintals* of bark, each of which can be sold for 300 Q. We have all experienced the corruption of local authorities. The previous mayor did not apply the law. He and his group promised many things but did nothing for the people of Totonicapan. Instead, some of them became millionaires overnight.

Many local leaders believe that the extent of corruption by high level authorities in Totonicapán has partly to do with the fact that they are generally 'outsiders', brought in from other departments by whichever party is in power. It was repeatedly put to us that the people filling such posts rarely understand, nor are particularly concerned about, local problems and issues.

To counter the activities of the bark strippers, certain communities have attempted to organize local vigilante groups but have run into serious difficulties. Many people are obviously reluctant to participate in activities which involve staying up most of the night; enduring extremely cold temperatures, wind and rain; and facing threats to physical safety. There is also a danger that police or army patrols will come across those participating in night-time vigilante activities and accuse them of being guerrillas. Several community leaders we spoke to recounted experiences of people being gaoled when found at night, while the real culprits went free.

The environmental and social impacts of such activities are very apparent. When a tree is stripped all the way around the trunk it dies since the flow of sap and water between the roots and the leaves is interrupted. Trees that are only partially stripped can survive but their growth is affected. Moreover, given that the bark acts as a protective shield against disease and pests, and cannot grow again, the tree becomes vulnerable to disease. The quality of the wood also declines as does the quantity which can be used for planks.

Bark stripping poses a threat not only to the white pine but to the pine forest in general. As the presence of this species (*Pinus ayacahuite*) is reduced, the coloured pine (*Pinus rudis*) comes to predominate. The latter is far more susceptible to attack by pine borer. Also, since these forests are generally located on steep inclines, it is probable that that the death of these trees will cause erosion and loss of soil fertility, thereby posing an obstacle to reforestation, natural regeneration and agricultural production.

Illegal bark stripping has also had a serious effect on the conservation activities of DIGEBOS and, in particular, the success of an imaginative project to promote communal forest protection and management. The project has attempted to encourage the local population in certain communal forest areas both to reforest and undertake a number of forest management practices – notably thinning – to reduce the threat of pine borer disease. Many local residents no longer want to participate in such schemes since it is apparent that reforestation and forest management practices have played into the hands of the bark strippers. Reforested areas along roadsides, for example, have provided the bark strippers with a readily accessible supply of bark. Thinning areas of forest to reduce the threat of pine borer disease has made access into dense forest areas far easier. In several communal forest areas people have lost all motivation to work in reforestation programmes and in the tree nurseries.

One community leader expressed his views on the subject as follows:

> Every year we participate in reforestation schemes. But now we are simply laughed at and told we are working for the bark strippers. In just a few minutes the people who strip the bark undo all that we have done in years. Twenty five years ago we began to reforest. We protected the seedlings from frost, covering them and uncovering them when necessary. Now those trees are slowly dying. Then we are told we cannot even use them since it is illegal to fell trees that are not dead.

What emerges from this discussion is that traditional communal forest protection mechanisms are breaking down as a result of illegal activities such as these and the fact that the rule of law has only limited application in relation to the Indian population. Indian land rights are not respected. Community-based protection mechanisms associated, for example, with patrols and the activities of forest rangers, as well as certain silvicultural practices are no longer practicable. A specific alliance involving tanneries, transporters, local labourers lured by high wages, and corrupt and repressive state authorities has served to alter the correlation of forces decidedly in favour of those who benefit from the destruction of forest resources.

Cultural perceptions which contributed to forest protection have, of course, been slowly modified throughout several centuries. The process of acculturation brought with it an alternative vision of the world and the universe. Rather than being an essential player in the life process, the forest was reduced to a material object, subordinate to the short-term needs of humankind. The ongoing weakening of traditional beliefs during the past two decades also has to do with the invasion of Evangelist sects which have rapidly set up shop throughout the Guatemalan countryside. These sects frown upon many traditional Mayan beliefs and religious practices which refer to the spirituality of nature.

The presence of two sharply contrasting cultural perceptions is very much in evidence in Totonicapán. Each has had a bearing on the way in which forest resources have been used and partially accounts for the differential rates of deforestation found in the communal forests and *parcialidades* on the one hand, and in private and municipal forests on the other.

Attitudes towards the forest have also changed with urbanization. As people of the *parcialidades* and communities leave rural areas to live in nearby towns, their relations with the forest are inevitably transformed. The daily 'rapport' with nature as a life-giving source (provider of fuel, fertilizer, water, etc.) is broken. Instead, people begin to see the forest as a source of products for commercial gain.

Communal forest protection mechanisms have also been weakened as a result of intra-community and inter-community conflicts associated with disputes over land rights. These have sometimes arisen when documentation regarding tenure status is incomplete or unclear. This can be illustrated by the experience of the following two cases.

One case involves the *parcialidad* of Sapon, located some 10 km east of the town of Totonicapán. An organization representing five other *parcialidades* in the department claim that, in legal terms, Sapon is not a *parcialidad* and that the administration of the forest should pass to the municipal authorities. They maintain that a group of people living in the area has merely used this status as an excuse to exploit the forest for themselves while the wider community has been denied access to forest resources. The families of Sapon claim that they do indeed posssess documentation proving that the forest is theirs and that the dispute has arisen because the municipality wants to extend its domain.

In the course of this study it was impossible to determine who was right. What was apparent, however, was that the dispute had served to weaken communal structures and forest protection mechanisms. When affected by conflicts such as these, the area in dispute can become the weak link in the chain of forest protection. As the dispute in Sapon unfolded, traditional forms of communal protection broke down and deforestation intensified. The negative impact manifests itself in two ways. Not only will people take advantage of this confused situation to exploit forest resources, but increasing social divisions prevent the Indian population from pulling together to protest about the more serious threats to the forest such as large-scale illegal commercial exploitation and the corruption of state authorities.

The other case is that of the community of Paqui, located approximately 5 km north-west of Totonicapán. In 1962 three *parcialidades* united to form a single community to strengthen their capacity to protect the forest which was being degraded by the activities of various families living in the area. In 1986, however, a group of families who were descendants of a past *cacique* and who were supported by the then president of the community association opposed a forest management plan which was to be implemented by DIGEBOS, claiming that this was simply an excuse to clear the forest for lumber. They attempted to seize legal documents governing property rights and initiatied judicial proceedings to claim the forest for themselves.

The case took five years to be resolved in favour of the existing community organization and the management plan was finally implemented. In the meantime, however, DIGEBOS refused to issue licences for tree felling. As a result the illegal exploitation of forest resources intensified, both by the group opposed to the plan and many other community members who depended on the use of forest products. Not only did the rate of illegal forest exploitation increase, but as increasing numbers of people were obliged to participate in illegal activities, people's perceptions of the forest and of forest protection changed. In such a situation, activities that are illegal and contrary to the usual norms of behaviour soon gain legitimacy and, from the perspective of many local residents, become an acceptable way of life.

'Illegality' may also prompt changes in the gender division of labour. Men, for example, have become increasingly involved in activities such as fuelwood gathering which traditionally have been the responsibility of women, children and the elderly. Women have been affected in different ways by the illegal exploitation of forest resources. In certain contexts where illegality leads to increasing conflict and violence, women may be displaced from traditional activities. In the *parcialidad* of Paqui, for example, we encountered an elderly lady who had abandoned fuelwood gathering in the communal forest near her home when the dispute over tenancy arose. Instead, she would leave the house at 6 am, walk several kilometres, gather fuelwood and return home at 6pm to sell a load (*carga*) of fuelwood for 3.50 Q ($0.70).

Other cases were encountered, however, where women formed an integral part of the clandestine network. Women would sometimes take charge of transporting fuelwood at night by cart to minimize the chances of being arrested. The police, apparently, were reluctant to detain women for fear of being accused of human rights abuses. The international campaign denouncing the human rights record of the Guatemalan government was, it seems, having some impact at the local level.

The historic divisions and mistrust which have characterized relations between Indian communities and state institutions are perpetuated and reinforced as people are drawn increasingly into a clandestine economy. This situation has affected the work of agencies such as DIGEBOS which need to win the trust of local communities in order to implement forest protection schemes.

Municipal and private forests

Significant areas of forest to the east and west of Totonicapán are formally classified as municipal forests. The origins of these forests probably date from the colonial era, when forest areas near towns were set aside as a source of wood for urban residents. Through time, there has been a tendency for municipal authorities to wrest control over communal forests, following a process similar to that, described above, currently affecting Sapon.

A permanent employee of the municipality takes charge of forest protection and liaises with the assistant mayors of the communities to organize the tasks associated with forest protection such as patrolling and reforestation. Moreover, a forest protection committee operates, with committee members holding their posts for two-year periods.

The municipalities can authorize the felling of trees within the urban area over which they have jurisdiction but must obtain licences from DIGEBOS to cut trees in rural areas. In practice, however, it is common for the municipal authorities to bypass the forestry agency.

While the failure to comply with the procedures laid down by DIGEBOS often implies less control over the use of forest resources, this is not always the case. In the municipality of San Francisco el Alto, for example, the authorities have taken the matter of forest administration entirely into their own hands. This situation arose out of a conflict with INAFOR, the predecessor of DIGEBOS. In 1978, the mayor was instructed by the Ministry of Agriculture to organize the reforestation of an area which had been partially affected by pine borer disease. Initially, the community participated actively in the scheme, forming a reforestation committee and establishing a large tree nursery. During the early 1980s, however, INAFOR cleared large areas of forest under the pretext that the trees had been affected by pine borer disease. The communities involved refused to collaborate further with the agency and set about administering the forests themselves. This situation has apparently yielded a number of positive results but prevented the new-look forestry corporation, DIGEBOS, from operating in the area.

Although a variety of forest protection mechanisms exist in municipal forests they are often less effective than in the case of the communal forests and the *parcialidades*. This is partly due to their proximity to larger urban centres and the sheer numbers of people who claim the right to use them for subsistence provisioning. Illegal tree felling often takes place in the municipal forests. The area currently experiencing the most rapid deforestation is the municipal forest of Panquix, just north of the town of Nahualá.

In addition, these forests are sometimes threatened by the actions of the municipal authorities who may, for example, turn over forest land for agricultural use. Such transfers of land may lack any sound socio-economic basis and simply constitute a 'deal' or favour beneficial to the local authority. Corruption has been rife in Totonicapán. Considerable animosity exists between the local population, who tend to regard the forests as their own, and the municipal and departmental authorities who commit abuses of power. In the run-up to Christmas, for example, local authorities have been known to cut and sell Christmas trees (*Pinus abeis*) which are classified as an endangered species. Tensions reached a head in the late 1980s when several local communities accused the mayor of Totonicapan and certain other authorities of cutting trees and selling timber for personal gain. A number of community leaders organized an inspection of the forest in question and approximately a thousand people visited the site. The mayor was taken to court and forced to resign.

Forest areas held as private property also cover extensive areas of Totonicapán. Many of these areas were legally constituted towards the end of the 18th century when the privatization of state lands took place and landowners were encouraged to obtain legal titles. Other private forests are of more recent origin, emerging when people from other departments such as Huehuetenango and Quiché migrated to Totonicapán and settled in more isolated forest areas where land titles did not exist. Once permanent residency was established, people would lay claim to the land. While it is difficult to generalize, there is some evidence to suggest that private forest areas have been subject to the most rapid deforestation as land has been converted to agricultural use or landowners have taken advantage of higher prices for wood and felled large numbers of trees. The regulations and social conventions which constrain deforestation in communal and municipal forests, either for clearing land for agricultural purposes or for felling trees, tend to be less strict in the case of private forest areas.

Also, landowners often lack the means to control the illegal exploitation of their forests. This is particularly so in the case of absentee landowners living in urban centres. Such forests are not only subject to illicit tree felling but also more intense sheep grazing and the associated risks of forest fires (set to encourage the growth of new grass) which impede natural regeneration of the forest.

Few private landowners appear to engage in reforestation and generally only plant trees as live fences or wind-breaks, or for ornamental purposes or fruit production. Few private owners take out licences as they believe the trees belong to them and as such they should be able to do what they want with them.

It is clearly not always the case, however, that private forests are in a worse state or more vulnerable to rapid deforestation than other types of forests. The land tenure system, social institutions of regulation and control and cultural attitudes are just some of the variables which explain different patterns of land use. Also crucial is the question of access to alternative employment opportunities.

Highly contrasting patterns of exploitation of private forests were found to exist, for example, in the municipality of Momostenango (see Map 5.2). In the parish of Panca, in the northern part of this area, private forests have experienced extensive degradation. Here, rainfall and soil fertility are low and extensive erosion has occurred. While the population has acccess to relatively extensive plots of agricultural land, they cannot produce enough to provide for their subsistence requirements. Direct cultivation of maize only provides for a few months' consumption. To obtain the money required for basic household provisioning many families resort to the collection and sale not only of fuelwood but also two products obtained from the pine species (*Pinus oocarpa*) which grows in the area. These products are *ocote* which are resinous pine splinters used to start cooking fires; and resin which is used to make *copal*, an incense-like substance used in Mayan religious ceremonies. Alternative employment opportunities in the area are extremely limited.

This situation contrasts with that of the private forests in the south of the municipality. While demographic pressure in the community of Santa Catarina is far greater, the private forests are in a reasonably good state. This is due primarily to the fact that pressure on forest resources has been reduced as a result of the existence of numerous alternative employment opportunities. In this area many people find work as weavers, carpenters, traders, masons and seamstresses.

THE CRISIS OF PETTY COMMODITY PRODUCTION AND SUBSISTENCE PROVISIONING

The change in the rate of exploitation of forest land over the past two decades is accounted for partly by the increasing demand for forest products and the fact that a significant sector of the local population must turn to the exploitation of forest products as a 'survival strategy'.

The increase in demand stems partly from sources outside the department. There appears to have been a qualitative change in the situation following the 1976 earthquake when Totonicapán became an important source of construction wood for other parts of the country. Municipal authorities came under considerable pressure from the central government to relax controls on the exploitation of forests and the transportion of wood to other departments.

Local consumption of forest products has also increased significantly. This is largely accounted for by the increased demand for fuelwood which is used by virtually all families in the department. The population of Totonicapán increased by 70 per cent between 1964 and 1990, from 170,000 inhabitants to 289,000 (including Nahualá) or roughly 58,000 families, each of whom uses, on average, a cubic metre of wood a month for fuelwood.

As indicated earlier, the bulk of the working population of Totonicapan and Nahualá are both peasants and artisans and/or traders. Unlike many other areas of the *Altiplano*, Totonicapán has been less directly affected by military conflict and forced population movements. Given, however, that the economic health of Totonicapán depended to a considerable extent on access to and the growth of markets in neighbouring departments, the crisis in the 'periphery' of the *Altiplano* has affected the well-being of 'core' areas such as Totonicapán (Smith, 1990).

This has affected, in particular, the livelihood of weavers, tailors and traders and, among these groups, led to processes of proletarianization, increased unemployment and underemployment, as well as out-migration to other regions or abroad. It is common for family members to migrate to work for several months in the cotton or coffee harvests towards the end of the year. There are indications, however, that employment and income-earning opportunities associated with these activities are declining. Several of these groups have also been affected by government economic policies and repeated devaluations of the Quetzal which have rapidly increased the cost of crucial inputs such as gasoline and thread. A similar situation affects carpenters but the reasons for the demise of the local carpentry industry, analysed below, are somewhat different.

Household expenditures in the area are extremely low. A fully employed labourer or piece-rate worker can expect to earn between 30 Q and 80 Q ($5–16) a week. Weekly household expenditure on food and a few other essential items varies from approximately 25–30 Q ($5–6) for a poor family, to 50–60 Q ($10–12) for those who are relatively better off. The local diet consists essentially of maize, eaten mainly in the form of maize-meal wrapped up and cooked in leaves.

By examining in some detail the situation regarding peasant agriculture, the local carpentry industry and the system of fuelwood provisioning, we can appreciate more clearly the linkages between deforestation and the crisis of traditional livelihood systems.

Peasant agriculture

The majority of rural families own agricultural land but plots are generally far too small to provide enough basic grain for household needs. Families grow principally maize for household consumption; beans and wheat may also be grown. In certain areas, it is common for families to have a small number of fruit trees. Historically, sheep grazing has been important although this activity, like that of wheat production, has declined in recent years. Families often own a few domestic animals such as chickens or pigs which are usually sold when additional income is required.

Maize yields generally vary from less than a quintal (100 lbs) per *cuerda* on unfertilized poorer soils to three or four quintals. Annual maize consumption per family is approximately 22 quintals (qq) per family.[5] A family obtaining yields of 2 qq/*cuerda* would need approximately 12 *cuerdas* (discounting for seed and some post-harvest losses) to be self-sufficient in maize. It is apparent that most families own insufficient land to meet even their basic maize requirements.

Historically, agriculture in much of the region has been characterized by the use of what can be considered fairly 'appropriate technologies' from the point of view of environmental sustainability, productivity and cost-efficiency. Traditionally, peasants have practised fairly sophisticated systems of crop rotation, or multi-cropping involving maize, potatoes, wheat, oats and broad beans. Terracing was common as was the use of organic fertilizers, including sheep droppings or hog dung mixed with straw and leaf-litter from the forest. As a result of such practices, maize yields in Totonicapan were somewhat higher than the national average, exceeding the latter by some 20 per cent in the late 1970s (Agricultural Basic Grains Survey, 1978).

While data on yield trends were difficult to come by there is some evidence to suggest that yields may have fallen slightly in recent years. Writing of the 1930s and 1940s, McBryde refers to yields of 1 to 2 quintals per *cuerda* with average yields nearer the higher figure (McBryde, 1969). During the mid-1970s, IIESO reported average maize yields of 1.23 qq/*cuerda* among producers with less than 1 *manzana* (approximately 70 per cent of the total) (IIESO, 1976; COGAAT, 1989). Among this group, 54 per cent obtained yields of between 0.5 and 1.25 qq/*cuerda*. Agricultural Censuses of the late 1970s indicate average maize yields of 1.3 and 1.4 qq/*cuerda* in 1977/78 and 1979 respectively.

During the past two decades, the sustainability of peasant farming systems has waned. In many areas the availability of forest leaf-litter has declined sharply. As a result farmers have been obliged to reduce fertilizer applications or switch to expensive agrochemical substitutes. A trade in leaf-litter does exist but it is generally only the more profitable commercial farmers of cash crops such as vegetables that can afford to purchase this commodity.

In many areas pressure on the land and the need to prioritize crop production has led to a reduction in sheep grazing or has meant that such activities now take place in areas which are distant from the plot. As a result, many crop or fallow areas no longer benefit from this natural form of fertilization. The spacial displacement of sheep grazing has generally been in the direction of the forests, though sheep and forests make poor partners in at least two respects. First the sheep trample seedlings and thereby restrict natural regeneration; and secondly, the forests are put at risk by the sheep graziers' practice of burning forest clearings before the onset of the rains to encourage grass to grow.

The opening up of new areas for crop land generally occurs on extremely steep hillsides which are prone to erosion and where soils are not suitable for agriculture. As distances between the household and the farm plot increase, and as more time is allocated to non-farm activities, labour intensive on-farm activities which could arrest erosion and/or enhance soil fertility such as terracing, multi-cropping, the installation of drainage systems, and the use of live and dead barriers (*barreras*) tend to suffer.

As plots are sub-divided through inheritance and yields decline, many families are forced to purchase an increasing amount of maize. Maize prices have risen sharply during the past decade, increasing faster than unskilled wages. Traditionally, many families grew wheat as a cash crop to generate part of the income needed to purchase basic food and other items. Wheat prices, however, have fallen relative to those of most other products (see Table 5.3 on page 79).

As household food requirements increase, multi-cropping or crop rotation have gradually been abandoned in favour of mono-cropping of maize. This has also occurred in response to the fall in prices of certain traditional cash crops such as wheat, particularly during the 1980s (see Table 5.3).

Another change in local farming systems concerns the shortening or virtual elimination of crop-fallow rotations. This is the result not only of increasing pressure on the land but also other factors. In certain high altitude areas, such as 'Alaska' in southern Totonicapan, it was common for peasants to sow crops for two or three seasons, and then turn the area over to pasture and sheep grazing for several years. By the fifth year the land was generally invaded by a coarse grass known as *pajón* unsuitable as pasture. Since the roots of *pajón* could be used for making brooms, it was not necessary to burn the entire area in preparation for sowing crops. Since the 1970s, however, there has been a sharp fall in the demand for 'natural' brooms (industrial brooms being purchased instead) and producers resort more frequently to burning which affects soil fertility negatively.

There have, apparently, been some perceptible changes in micro-climate over the past several decades which have affected agriculture. While it was not possible to verify this with statistical data, many of the older peasant producers and community leaders consulted during the course of this study claimed that there had occurred a significant reduction in rainfall over a 30- to 40-year period. A liberal translation of the words of one village elder depicts what many feel to be the situation: 'Guatemala is known as the land of eternal spring but the spring is now drying up'.

It is important to note the tremendous concern which exists in the area about the impact of deforestation on the water regime. In meetings held with community leaders and forest protection committees to discuss the impact of deforestation on the lives the local population, this theme was stressed repeatedly: 'Our wells and natural water-holes are slowly drying up', was the standard comment. This problem has become particularly evident during the past two decades and is of far greater concern than other ecological problems associated, for example, with erosion or the loss of biodiversity. The reason for so much concern regarding natural water sources is indicated by the statistics on water usage. According to national census data, nearly half the households (48 per cent) of the department of Totonicapan depended on wells (32 per cent) or natural sources such as springs or rivers (16 per cent) during the early 1980s. Many living in upland watershed areas were aware that deforestation was affecting the water supplies of populations in lowland areas, including the town of Totonicapan.

The local carpentry industry

The decline in employment and income-earning opportunities is particularly noticeable in the local carpentry industry. The several thousand carpenters operating in Totonicapan and Nahualá are finding it increasingly difficult to make ends meet. Production costs have increased sharply over the past decade while supplies of good-sized wooden boards are dwindling as the larger trees disappear. This situation has given rise to a fairly intense process of social differentiation whereby poorer carpenters are abandoning self-employment to work on a piece-rate basis for larger carpenters.

One carpenter who was interviewed claimed that when he was self-employed his profit margin per wardrobe was as little as one dollar (a wardrobe usually takes between 1.5 to 2 days to make). Three years ago he had been obliged by economic circumstances to take on paid work. He now earns 60 Q ($12) a week making large wardrobes for a local government official – barely enough to purchase the family's basic food requirements.

Several larger carpenters have been able to obtain loans from banks or private foundations. This was the case of one carpenter interviewed in the valley of Totonicapan. Prior to 1983, he had worked as a pit sawyer in the mountains where he would spend periods of 15 days as one of a team of workers employed by someone else. Conditions were extremely harsh and towards the end of the 15-day period it was common for the men to go hungry and become ill. He began to make small wardrobes and in 1986 was able to obtain a bank loan which enabled him to obtain a stock of wood, improve his house and buy land. He purchased several *cuerdas* of land in the valley and another 12 in the mountains. He now owns 25 *cuerdas* and produces enough maize to feed a family of eight all year round.

He also began to employ carpenters and now has four men working for him from their own homes, earning between 25 Q and 30 Q ($5–6) per wardrobe. Each worker makes approximately 12 wardrobes a month. Selling each wardrobe for 150 Q ($30), the owner made a net profit of approximately 25 Q ($5) per item which gave him a weekly income of about 300 Q ($60), considerably more than the weekly income of 30–60 Q ($6–12) earned by most fully employed labourers in the region or 75 Q ($15) earned by subcontracted carpenters.

He was of the opinion, however, that the future of the local carpentry industry was bleak. More and more self-employed carpenters were seeking work on a piece-rate basis. It was becoming increasingly difficult to purchase wood. Much of the wood was now leaving the region for other departments and the major cities. Deforestation in the area during the past decade had meant that the boards came in ever-decreasing sizes. A few years ago it was easy to find boards of 12 to 14 inches in width. Now they were generally 7 to 9 inches wide and getting narrower. He now needed 15 boards to make a wardrobe, compared to 14 the previous year and 12 five years ago. These developments had had the effect of increasing his costs of production.

While this carpenter was clearly in a better position than most, he saw little prospect for remaining in the carpentry business. His strategy for the future was to carry on producing furniture for just three to four more years. 'The trees are disappearing and those that remain are very small. It is already difficult to get planks and fuelwood. What is more, the people of the communities no longer want to see the trees cut down.' He was already making plans to teach his sons how to weave and then set up a small weaving business as an alternative to carpentry. It was also likely that with sufficient capital, he would purchase a trade vehicle and engage in commercial activities.

The days of the small self-employed carpenter in Totonicapan seem to be numbered. Costs of production have increased substantially in recent years and the rise in the price of items such as wardrobes is clearly having a negative impact on demand. Carpenters now either have to work on a piece-rate basis or, to be successful, have a working capital of around 5,000 Q ($1,000). A month's stock of wood alone for a carpentry enterprise of this size would cost around 3,000 Q. Data from the Guatemala Foundation on costs of produc-

Table 5.2 Costs of production and revenue in a carpentry enterprise, 1989

Costs	**Q**	**%**
Labour	800	18
Wood	2880	64
Transport	85	2
Other	703	16
Total	4468	100
Revenue	5280	
Profit	812	(14%)

Source: Fundación Guatemala.

tion of local carpenters indicate that wood makes up about two-thirds of total costs (see Table 5.2).

Data for late 1989 show the following breakdown for one fairly typical enterprise producing 32 large wardrobes a month with four workers (including one family member). It should be remembered that by 1991 the costs of production had probably increased by between 30 per cent and 40 per cent.

Very few carpenters can amass 5,000 Q. While a number of banks and foundations provide loans, these are generally restricted to those whose activities could be classified as small businesses (*micro-empresas*) and meet certain conditions, namely, that they own property of a certain value and utilize a simple accounting system (which implies a certain level of education). One of the main organizations granting loans to small businessmen was the Fundación Guatemala. During 1989 and 1990, however, only 59 carpenters in the region had received loans, usually of between 1,500 Q and 2,500 Q ($300–500) from this organization.

Certain other carpenters in the valley of Totonicapan who resisted becoming piece-rate workers were considering other options, notably trying to produce finer quality furniture for a more middle-range consumer market. Various strategies were in evidence or were at least being considered. Some carpenters talked of obtaining hardwoods such as cedar and mahogany from the southern coast and of changing designs to produce finer quality furniture. Others were trying to market their own products directly on the roadside or in nearby towns.

Carpenters in Nahualá, for example, with the assistance of the Guatemalan NGO, FUNDAP, engaged in a strategy of adding value to what they produced by diversifying production to incorporate items such as Indian masks and other decorative items for the tourist market. The notion of producing more 'up-market' goods was not new to the carpenters of Nahualá. Some 30 years ago local merchants from the town began buying pine furniture from carpenters in Totonicapan. Before selling the furniture, however, they would leave it for several weeks to hang from the ceiling of their houses above an open-hearth fire. This smoking process gave the furniture an antique look and sales increased, particularly in the capital, Guatemala City. To meet the demand, local production in the community of Nahualá increased. Today there are some 200 carpenters. Because the smoking process took so long, carpenters switched to using synthetic dyes, such as shoe polish, to produce the same effect.

The area in and around Nahualá has been largely cleared of trees and it is becoming increasingly difficult to obtain the white pine used by the carpenters. This has put tremendous pressure on the nearby forest of Panquix – where much of the region's illegal timber extraction is centred – to supply the valley of Totonicapan and Nahualá as well as cities like Guatemala and Chimaltenango.

Fuelwood provisioning

Another important element in the crisis of subsistence provisioning concerns the increasing scarcity of fuelwood. The importance of fuelwood in the local economy is immediately apparent when driving through the region, be it on the main roads or off the beaten track. One encounters a constant stream of people of all ages walking along trails or by the roadside, hunched over, with heavy packs of fuelwood on their backs.

For the Mayan Indians of the *Altiplano*, living at high altitudes, fuelwood assumes a dual purpose. It is not just cooking fuel but is also a source of heat. Apart from considerations of cost, this is one of the reasons why few households have made the transition to gas or electricity for cooking or have switched to improved fuelwood stoves. An estimated 94 per cent of all households in the region use fuelwood. The growth in population, of 120,000 between 1964 and 1990, has put tremendous pressure on fuelwood supplies.

Until fairly recently, fuelwood gathering was not a major cause of deforestation. Most communities and *parcialidades* imposed fairly strict regulations governing tree cutting and permitted local families to cut branches or use dead wood and twigs. Rather than cut trees, many fuelwood gatherers would scale the trees and remove all but the upper branches.

While considerable quantities of fuelwood are still gathered in this manner, many families now resort to the occasional felling of a tree. By law such activities require the permission of DIGEBOS but, as indicated earlier, few families acquire the necessary permits. Tree felling for fuelwood provisioning, therefore, generally assumes a clandestine character. Control of such activities has given rise to numerous conflicts between the local population and state authorities who at times harass, confiscate, fine or even imprison the people involved. This has generated much resentment among the local population both because such activities are regarded as necessary for survival and because those agents – usually outsiders – who exploit the forest illegally for commercial gain often operate unimpeded by the police or can simply bribe their way out of any encounter with the authorities.

The increasing scarcity of fuelwood has had a dramatic impact on the household economy, everyday work routines and the gender division of labour. As mentioned above, the average household consumes approximately one *tarea*, roughly a cubic metre, of fuelwood a month. Traditionally, women (particularly the elderly) and children have been responsible for collecting fuelwood on a daily basis. Sometimes this activity would be combined with shepherding sheep or children would dedicate time to fuelwood gathering after school.

During the past 10 to 15 years there have been three fundamental changes in the system of fuelwood provisioning. These are that the distances travelled have increased considerably, the family division of labour has changed, and increasing amounts of fuelwood are now purchased.

Several families in rural communities reported that whereas five years ago they would spend three hours to half a day per week collecting fuelwood, they now spent between half a day and a day. As one local community resident observed:

> In many areas it used to be easy to gather fuelwood since the communal forest was nearby. But several people who had no land asked the community leaders (*alcaldes auxiliares*) for land. They would be given 2 to 3 *cuerdas* of forest land and bit by bit the communal forest was parcelled off. Now it takes us half a day to collect fuelwood.

While distances travelled vary considerably, of course, they generally range from between 1 and 5 km for people in communities located near forest areas to as much as 16 km for certain families living in the densely populated valley of Totonicapan.

Families in many areas have had to radically change the way in which they organize fuelwood collection. It is common to find those who dedicate an entire day each week to fuelwood collection, with all the family leaving the house early in the morning to return at dusk. Increasingly, men have been drawn into this activity so that larger loads can be brought home. Men generally carry a hundredweight or more, while women and children carry between 25 and 50 lbs. Even small children will be found carrying their 10 pound loads. Families are obliged more and more to use fuelwood substitutes such as pine cones or *olote* (the maize cob) which previously was used as fertilizer.

For many families fuelwood collection is no longer a casual spare-time activity undertaken by the elderly and the young. It has now become hard work undertaken by all the family. In certain areas the point has now been reached when families begin to purchase fuelwood or, as is more likely, a proportion of their fuelwood requirements. The income earners of the household have to weigh up carefully the relative advantages of allocating a day to fuelwood collection or earning a day's income. The impact on the family budget of having to resort to fuelwood purchases can be dramatic. Purchasing a month's supply of fuelwood costs approximately 60 Q ($12) or up to 80 Q ($16) if purchased in small bundles (*cargas*), one at a time. This represents the equivalent of one to two weeks' full-time work.

The rise in fuelwood prices has been greater than that of wages. Whereas the price of a *tarea* and a *carga* of fuelwood rose 500 per cent and 567 per cent, respectively, in just over a decade (see Table 5.3), the basic agricultural wage rose 300 per cent in nominal terms.

Table 5.3 Relative prices of selected forest products and other items in Totonicapán, 1980–91 (quetzales)

Commodity	**1980**	**1986**	**1990**	**1991**	**1991/80 Index**[1]
Forest products:					
Tree ('medium-size')	12	–	–	100	833
Dozen Planks:					
– 12 inch[2]	12	60	–	120	1000
– 7 inch[2]	10	48	60	70	700
– Purchased at site[3]	3			50	1667
– Planed[4]	–	75	100	180	–
– Rough[4]	–	40	60	80	–
Fuelwood:					
– *Tarea* Pine[5]	10	32	44	60	600
– *Carga*[6]	0.60	–	–	4	667
– Transport[7]	–	70	100	250	–
Inputs and food:					
Gasoline (Regular)	2.18	3.35	4.60	8.95	411
Diesel	0.72	1.76	2.80	5.95	826
Fertilizer (100lbs)	12	20	50	71	592
Maize (100lbs)	5	–	–	40	800
Products sold by petty commodity producers:					
Wardrobe (large)	12	35	105	150	1250
Wheat (100lbs)	14	18	–	35	250
Labour power:					
– Daily wage[8]	2	–	–	8	400
Dollar	1		4	5	500

1. Base year 1980 = 100.
2. Prices reported by carpenters.
3. Purchased in the forest.
4. Prices reported by lumber merchant near the town of Totonicapan.
5. One *tarea* is roughly equivalent to 1 cubic metre.
6. 20 *cargas* = 1 *tarea*.
7. 15–20 km trip by truck.
8. Basic wage in agriculture.

This fact, combined with the increased rates of unemployment and underemployment, has forced families in certain areas to rely increasingly on fuelwood gathering as a source of income. This is most noticeable among the poorest families in areas such as the north of Momostenango, Santa Maria de Chiquimula and Santa Lucia de la Reforma where alternative employment opportunities are minimal. In these areas agricultural productivity is extremely low and many families have become increasingly dependent on the sale of forest products.

A fuelwood merchant near the town of Totonicapan, interviewed for this study, explained some of the changes which had occurred in the trade during the past decade. Fuelwood was coming increasingly from private forest lands. A truck owner would buy several trees, pay one labourer to fell them and cut the trunks and branches into fuelwood, and another to load the truck. Increasing controls on transporters were making it more difficult to obtain supplies: many refused to pay the costs of licences and permits or to lose the time needed to obtain the necessary papers. They preferred to risk their luck, either avoiding roadside controls or hoping that a bribe would work out cheaper. This forced several operators to work at night when controls were virtually non-existent. There were some indications that profit margins of transporters had fallen. Whereas fuelwood prices had risen by approximately 100 per cent during the past five years, gasoline prices had increased by about 200 per cent. It is probable, however, that the wages of labourers working for transporters had been squeezed to safeguard margins. Piece rates of as low as 1 Q to 3 Q (20–60 US cents) were reported in relation to work associated with the cutting and loading of wood from one tree.

Some transporters were apparently withdrawing from the fuelwood trade. To fill their place, smaller traders would bring fuelwood from distances of 10 to 16 km, on foot or by mule. Generally, though, supplies were less reliable. This, combined with the fact that personal relations with fellow community residents were somewhat strained as a result of ever-increasing fuelwood prices, had led the merchant to consider withdrawing from the fuelwood trade and expanding other activities, notably irrigated vegetable production.

Changes had also occurred in the quality of the fuelwood which was traded. Slow-burning fuelwood from hardwoods such as oak (*encino* – Quercus) was now in short supply and, depending on the area, approximately 40 per cent to 100 per cent more expensive than pine. Logs were now smaller, the wood was of a poorer quality and loads contained many small branches. Complaints regarding quality permeated the entire fuelwood chain. Consumers also claimed that the standard measure used to sell fuelwood, the *tarea*, was no longer a genuine *tarea*. Several fuelwood merchants had apparently reduced the width of the *tarea* from 33 to 30 inches.

The increase in fuelwood prices had meant that most families could no longer buy in bulk. This, in effect, meant that the price of fuelwood for poorer families was approximately 50 per cent higher than for families which could purchase a month's supply at one time. Whereas the profit margin on a *tarea* of fuelwood was around 10 per cent, that obtained when the same volume of wood was sold in small bundles increased to more than 60 per cent. A merchant interviewed for this study would buy a truck-load a month (10 *tareas*) and could expect to earn a profit of approximately 300 Q ($60). The fuelwood trade was just one of several activities in which he engaged. He was also a tailor and a member of a co-operative producing vegetables.

The case study of Totonicapan revealed that increased deforestation has resulted from a complex set of conditions associated largely with the breakdown of traditional resource management systems. The system of communal protection mechanisms has been weakened considerably, with communal regulations having been gradually replaced by a code of state regulations which remain largely unenforced. Community structures and traditional regulatory mechanisms prove ineffective when confronted with the growth of a clandestine economy, centred on the exploitation of forest resources, and associated aspects such as violence and night-time activities. They also prove ineffective when confronted by a state apparatus which fails to guarantee community or customary rights, discriminates against Indian communities in the application of the rule of law, and aids and abets those engaged in the organized illegal exploitation of forest resources.

Pressures on the natural resource base have intensified, partly due to population growth. More importantly, however, this is a result of the crisis of the traditional economic system characterized by multiple employment opportunities, and the fact that circuits of surplus extraction largely benefited the local population. Since the 1970s, the local economy has been affected by recession and changes in processes of surplus extraction whereby value added is appropriated increasingly by external agents.

From the above discussion, it is evident that deforestation has had a major impact on the lives of the population of Totonicapan. The analysis of the situation of carpenters and fuelwood gatherers revealed the extent to which livelihood, social relations of production and the gender division of labour have been affected.

Important changes in social relations have taken place within the local carpentry industry. As the cost of boards has increased, many have been forced to abandon self-employment and either give up their trade or find employment as paid hands of larger carpenters. The process of social differentiation among carpenters has intensified during the past decade. The availability of credit from certain financial and aid agencies has enabled a small group of carpenters to expand their businesses and employ non-family labour.

During the past 10 to 15 years there have been several fundamental changes in the system of fuelwood provisioning. The distances travelled and the time allocated to gather fuelwood have increased considerably. Fuelwood gathering has become a much more burdensome activity. This in turn has prompted changes in the gender division of labour within the household as men become more involved in this activity. Increasing amounts of fuelwood are now purchased. This change can have a significant impact on the household budget.

An important phenomenon asociated with deforestation is that of 'illegality'. Increasing numbers of people have been drawn into a clandestine economy centred on the exploitation of forest products. This has served to further weaken communal forest protection mechanisms and has had a major impact on lifestyles, community relations and relations with the state and development agencies.

Chapter 6

Deforestation and shifting peasant agriculture

For centuries, peasant agriculture in Central America has been based on the slash and burn or *roza* system. Such a system can yield important social, economic and ecological benefits under conditions where the person to land ratio and the cash and consumer demands of the peasant household are low (Heckadon, 1982). It not only provided peasant families with their basic food requirements but also minimized risk, given that few or no costly modern inputs were required and indebtedness was restricted. By burning the dense covering vegetation of forest areas the peasant transformed this biomass into nutrient-rich ashes which fertilized crops. The *roza* system, however, depended on the possibility of leaving land fallow for periods of 10, 20 or more years as it was impossible to continue production on the same plot under conditions of declining fertility and prolific weed growth. The natural regeneration of forest served to increase the fertility of soils, improve their structure, protect them from erosion and reduce the incidence of weeds (CIERA, 1984:83–7; Heckadon, 1984:218).

In many wet tropical forest areas soils tend to become quickly exhausted. Not only are such soils generally poor in nutrients, but also much of the ash deposits derived from burning are lost as a result of leaching and wind and rain erosion, while the types of grains grown by peasant producers are generally highly nutrient-demanding (ibid.:217). Moreover, the producer faces the problem of rapid weed growth which tends to intensify the longer the plot is used. A point is reached when an excessive amount of labour has to be invested in production while there is a tendency for yields to decline. Traditionally, the producer would clear, burn and sow another plot, leaving the previous plot fallow for several years before returning.

Such a system, however, is extremely fragile and liable to break down as pressure on the land and the rate of deforestation increase. When this occurs, the impact on agricultural production and productivity, employment, livelihood and social relations can be dramatic.

DEFORESTATION AND LAND COLONIZATION IN PANAMA

These relationships have been analysed in some depth by Heckadon in a study of colonization processes in Panama (Heckadon, 1982, 1984, 1985). As access to forested land becomes more restricted and the demands of the peasant family for food and cash income increase, the producer generally opts out of the slash and burn or *roza* system.

> In colonization frontiers, the disappearance of forests occurs in two ways. Initially, only a small number of farms, usually smaller ones, exhaust their fresh supplies. But plots can still be found in larger farms, those of wealthier persons who as a rule want to seed the plot with pasture. In this first instance the scarcity of forest is limited or localized to a section of the frontier. Finally there comes a second stage in the evolution of the relationship between man and the forest, then this resource is completely exhausted ... *'cuando se acaban los montes'*, when there is no more forest to farm, and the crisis of *roza* agriculture becomes generalized. (Heckadon, 1984:221)

To understand the conditions which lead to the breakdown of *roza* or swidden farming systems in Central America it is crucial to refer to the effects of cattle expansion on peasant farmers. As indicated in Chapter 2, the cattle boom from the 1950s onwards not only saw large tracts of primary forest converted to pasture but also resulted in the reduction of 'idle' or fallow land which formed an essential part of the *roza* system. Land which peasants had previously rented or been allowed to farm as a 'favour' for labour services suddenly became a valuable commodity beyond the reach of most peasant farmers (Brüggemann and Salas, 1992). Moreover, as noted in a study of changes in farming systems in Costa Rica, cattle often destroyed the crops of peasants who could not afford to fence their holdings while newly introduced grasses dispersed rapidly, sometimes competing with food crops (ibid.).

The response of peasant producers to such pressures varied considerably. Throughout the region, many have been forced to sell their plots (or improvements). Heckadon describes a process whereby peasants in the case study area often converted their land to pasture for cattle grazing in the belief that this activity would provide greater cash income (Heckadon, 1984:219). But many small producers who took up cattle raising soon embarked on a cycle of economic decline.

> .... [S]oil erosion, resulting from the summer burning of fields, overgrazing, and the mechanical action of the cattle moving up and down the slopes, becomes a serious problem. Weed control becomes a costly struggle which, after a time, the majority of the small and medium-sized producers cannot endure. The cattle grow slowly, and what is worse, productivity, like income, gradually declines. Moreover, the marketing system operates against the small producer since the greatest profits are obtained by large-scale producers and above all by middlemen.... In the cattle zones people consume less meat and milk than they did a generation ago. When the families can no longer live from cattle raising, they sell their farms. In this way a process of land concentration begins and people migrate to ever distant frontiers. As productivity declines, unemployment increases and land concentration occurs, a new a more unstable social order takes shape which potentially is quite explosive. (Heckadon, 1985:50)

Social relations of production have been radically transformed with deforestation and the breakdown of the *roza* system. Members of peasant families are often obliged to engage in sharecropping as a means of obtaining additional plots of forest land or must seek employment as wage labourers. It is often the case, however, that the dynamic sector in the local rural economy is cattle raising which provides relatively few employment opportunities.

This is brought out clearly in a study of the cattle industry in Nicaragua and Honduras where the employment ratio per hectare of pasture during the early 1980s was approximately five or six person-days per annum. This compared with 202 person-days in the case of coffee and 73 for maize in Honduras (Howard, 1987b). The upshot of this situation was that the cattle boom which occurred in several Central American countries generally led to intense processes of out-migration to urban or agrarian frontier regions. As Heckadon explains:

> The extinction of the forest, followed by serious ecological deterioration, has negative effects on ... social organization; there is a breakdown in the system of subsistence agriculture and a phasing out of all traditional institutions of labour exchange and credit based on mutual trust. As the productivity of ranching decreases, and land impoverishment increases, family farms are unable to generate sufficient income. Many of the smaller and medium-sized landowning *campesino* settlers begin to sell their land to larger producers. As the land tenure system becomes more unequal, it leads to greater social differentiation among the peasants themselves. Moreover, when in the frontier most lands have been transformed to pastures for cattle, unemployment and underemployment become serious social problems. (Heckadon, 1984:221 and 277)

MIGRATORY AGRICULTURE IN NICARAGUA

The causes and impacts of the crisis of peasant agriculture in a zone in the northern interior of Nicaragua known as The Segovias was the subject of a study by the Nicaraguan research centre CIERA (CIERA, 1984). At the beginning of this century, a family could produce sufficient grains from just 1.5 to 2 ha to meet household consumption requirements. Maize yields probably exceeded 28 qq per ha. If we assume a 20-year crop–fallow rotation, then each family needed between 29 and 44 ha to meet its minimum grain requirements. The land–family ratio at that time was approximately 73 ha. By 1980, the same ratio had fallen to 13 ha (ibid.:89). Land concentration, however, was such that by 1980 just 300 families controlled approximately half the agricultural land while the other half was in the hands of 8,000 peasant families. The impossibility of maintaining the *roza* system had a dramatic effect on yields. Maize yields of 18 qq per ha were reported in both 1963 and 1974 (ibid.:242). Several peasants interviewed reported falls in yields of 50 per cent or more over a generation.

What was once a fairly uniform social structure, comprising peasant farmers engaged in slash and burn subsistence agriculture, became increasingly differentiated. A few rich peasant producers, many of whom were also merchants, accumulated land and capital and joined the ranks of incoming capitalist producers engaged primarily in agro-export production. The living levels of the vast mass of peasants declined steadily. As land became increasingly scarce peasant families no longer were able to produce the food they required. Diets became extremely monotonous and devoid of all animal protein sources. One study conducted in 1977 found that 70 per cent of the infant population in rural areas suffered from malnutrition. Many peasant families were forced to farm in increasingly inhospitable conditions, often on hillsides, where the introduction of ploughing and animal traction increased erosion and reduced yields. In certain areas (such as Cusmapa) erosion had led to desertification. Most peasants ended up selling their labour on a part-time basis in the coffee and cotton harvests and remaining unemployed for part of the year. Many also engaged in sharecropping. Others migrated to the agrarian frontier or to the towns and cities in search of better opportunities (CIERA, 1984:18–19, 321, 327–34).

In some of the more mountainous areas of The Segovias, the traditional *roza* system continued to operate with a reduced fallow period. Only small quantities of food could be produced by peasants using this system. Whereas in previous decades many of these producers, like their forebears, would have worked as wage labourers for short periods of time as a secondary activity, now they worked primarily as wage labourers and complemented this activity with direct production in the mountains.

As indicated above, one of the most pervasive socio-economic forces underlying deforestation in Central America relates to processes of colonization of rainforest areas. The introduction of technological practices, including ploughing and permanent cropping, unsuited to certain rainforest environments, has had serious effects on agricultural productivity and the livelihood of peasant families who inhabit such areas.

When analysing the question of social impact it is important to distinguish between the situation of 'spontaneous' settlers and those participating in organized settlement schemes. Some studies have found that, contrary to what might be expected, the environmental and social consequences of colonization have been worse in the case of government-sponsored settlement schemes than in the case of spontaneous colonization.

This is illustrated in a study of colonization in the central part of Nicaragua's Atlantic Coast region (CIERA, 1981):

> While the process began fairly recently (approximately 30 years ago) and has accelerated in terms of the increasing numbers of inhabitants one finds in the area, it has nevertheless been gradual. This fact has enabled the families involved to minimize the risk of failure. Immigrants take advantage of family ties and friendships, they learn the secrets of the mountains and land from those who arrived previously (... where to find the best plots of land, where the river floods, etc.) while the head of the family usually finds an appropriate spot before the rest of the family arrives. This process facilitates not only adaptation to the natural environment but also social and economic integration....
>
> In contrast to this situation are the numerous experiences of planned colonization in many Latin American countries, which have failed dismally. In such contexts colonization is generally quite rapid with the bulk of the immigrant population arriving at the same place at the same time. In such circumstances it is not possible to minimize the risks through family and friendship networks. The transmission of knowledge regarding the new ecological conditions is highly restricted ... while the decisions taken regarding production are often deter-

mined by the guidelines laid down by government planning agencies. (CIERA, 1981:259–260)

In Nicaragua's Río San Juan region, in the southern-interior part of the country, a case study was undertaken to examine in more detail the situation of peasant producers engaged in migratory agriculture in agrarian frontier regions.[1] One of the principal purposes of the research was to examine the effects of radical structural change, namely, agrarian reform and population resettlement, on deforestation and the livelihood of peasant producers. This aspect of the study is reported on in more detail in Part III of this book.

The findings of the Río San Juan study confirm many of the observations already mentioned regarding the vulnerability of peasant agriculture and processes of social differentiation and pauperization. They also highlight the difficulties families encounter in establishing a permanent base. During the 1960s and 1970s waves of colonists arrived in Río San Juan, principally from the northern- and southern-Pacific regions of the country. Many had been farm-workers or artisans and lacked a peasant identity or culture. Those from the dryer Pacific-coastal regions with some agricultural experience generally knew nothing of conditions in tropical rainforest areas. Although soils and climate were generally inappropriate for basic food crop production, their efforts focused almost exclusively on the cultivation of maize, rice and beans. Average maize yields rarely exceeded 14 *quintales* per hectare, approximately half the national average. Knowledge of permanent crops and agroforestry or silvi-pastoral systems was virtually non-existent. Considerable environmental degradation was caused by opening up land for agricultural purposes along river beds and in watershed areas.

Peasant livelihood systems were characterized by both a high degree of mobility and provisionality. Many colonists would settle on a temporary basis in areas controlled by large graziers/*latifundistas*. They would clear the forest, sow grains for one or two cycles and leave the land seeded for pasture for the rancher, before moving on to another plot. Other colonists would lay claim to areas of approximately 50 *manzanas* (35 ha) of forest land, clear approximately two to three *manzanas* a year and sow no more than five *manzanas* with grains. Following two or three seasons, the land would be converted to pasture and eventually sold to graziers. The sale of these *mejoras* or 'improvements' constituted one of the principal sources of cash income for the peasant family and was crucial for ensuring that the peasant household could continue to reproduce itself on the basis of direct production of the land and not via proletarianization or migration to urban areas.

Even those peasant producers who established a permanent base would tend to use the land they owned for cattle raising and continue to clear forest in the hills to plant grain. Considerable time would therefore be spent away from home.

This system benefited primarily a small group of graziers and merchants. Over a period of roughly 20 years, a few peasant households were able to acquire larger holdings of around 100 *manzanas*, and diversify their production by introducing both cattle and a greater variety of crops. Even these producers, however, would continue to sow grains using the slash and burn system in areas outside their farms. These instances of 'upward mobility' appear to be related to a combination of often extraordinary conditions: for example, differential rent obtained by early colonists who acquired relatively fertile land; close proximity to roads and markets, the existence of a relatively large number of 'family hands' which meant that larger areas of forest could be cleared more quickly; and the possibility that certain family members could find off-farm employment and thereby obtain income for both subsistence provisioning and accumulation (purchase of cattle, more land etc.).

In general, however, conditions for the bulk of the peasant population deteriorated. This was partly accounted for by the geographical trap associated with migratory agriculture here and in many other rainforest areas of Central America. As indicated in Chapter 2, the migration patterns generally associated with shifting agriculture are often in the direction of areas even less appropriate for basic grains production and cattle raising. As agro-ecological conditions deteriorate and/or as producers submit to the economic necessity to sell land 'improvements', peasant families move on, generally in an eastward direction where levels of rainfall increase significantly (see Map 6.1) and where soil conditions are often worse. Not only are yields affected but post-harvest losses, which can account for over a third of the grain crop, increase significantly due to difficulties in marketing, drying and storing grain (CIERA, 1989). Living levels for the more disadvantaged groups are apt to deteriorate, as a result of adverse production conditions, the increased risk to health and the fact that access to social services is more restricted.

By the time of the Sandinista revolution, social conditions in Río San Juan were amongst the worst in the country. The department recorded the highest illiteracy rate, at 96 per cent of the population. Given the mobile and provisional character of life, houses in rural areas lacked any solid structure. In fact, it is reported that before the revolution in 1979 local officials and certain graziers prohibited the construction of permanent dwellings with walls made of boards so as to prevent human settlement patterns that would obstruct the expansion of grazing lands for cattle. Whereas approximately 30 *latifundia* controlled approximately 150,000 ha, some 3,850 peasant families possessed 30,000 ha. Seventy per cent of the peasant population lived in fairly remote areas. Many producers with 35 ha or less had to find work as wage labourers for several months of the year in order to survive. Others were often obliged to migrate to urban areas or other regions of the country.

The policy and structural changes introduced by the Sandinista government following the overthrow of Somoza in 1979 were to alter quite dramatically the system of migratory agriculture in Río San Juan. The environmental and social impacts of these developments are examined in Chapter 11.

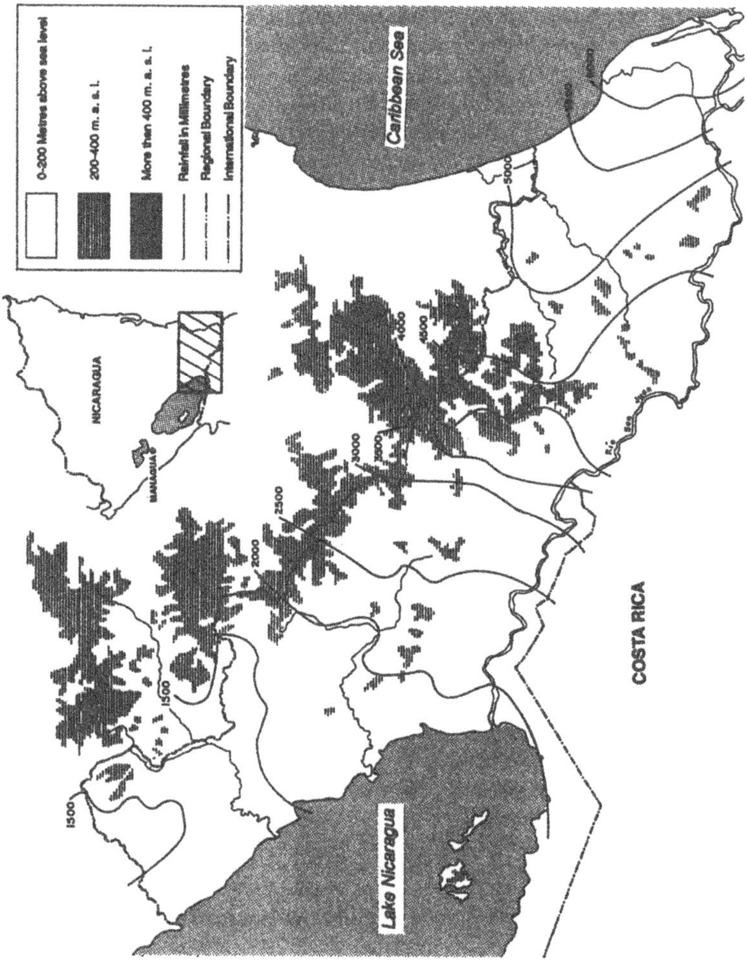

Map 6.1 Rainfall and topography in Río San Juan, Nicaragua

PART III
FOREST PROTECTION AND TREE PLANTING INITIATIVES

Overview

Both public and private concern regarding deforestation and other environmental problems have heightened dramatically throughout the region during the past decade. As a result, a number of important conservation initiatives involving forest protection and tree planting have been taken in most countries. As Budowski has pointed out, it is no mean feat that steps have been taken along the path of sustainable development in the midst of armed conflict and economic and social crisis (Budowski G., 1992).

Since the early 1970s, there has been a major increase in the number of protected wildland areas in the region. These include several international 'peace parks' spanning two or more national borders. Agroforestry research and dissemination have been actively promoted by numerous international and national agencies. A number of countries have introduced incentive schemes to promote reforestation and tree plantations. With the exception of El Salvador, governments in the region had completed final or draft Tropical Forestry Action Plans by late 1991. Natural resource policy inventories, assessing the effectiveness of environmental policies and programmes, had also been prepared for Guatemala, Honduras, Panama and Costa Rica.

At the regional level, there has, in recent years, been a considerable flurry of presidential (or First Lady) activity centred on environmental issues. In 1989, this led to the establishment of the Central American Commission on Environment and Development (CCAD) to promote regional co-operation to protect and rehabilitate the environment (Arias and Nations, 1992). These developments have served to intensify government and NGO efforts in this field and attract increasing amounts of foreign aid to support conservation projects and programmes. The levels of aid now available for environmental protection projects and programmes will no doubt mean that the scale and scope of forest protection schemes will expand considerably into the 1990s.

Non-governmental conservation organizations and environmental social movements have grown in strength during the past two decades and, in most countries, now constitute influential pressure groups (Annis, 1992a, 1992b; Cox, 1992; Fournier, 1991). Moreover, certain non-governmental organizations in countries such Panama, Costa Rica and Guatemala administer major forest protection schemes. The approach of such groups towards environmental and development issues, however, varies considerably. One, which may be labelled 'conservationist' and which appears to prevail in countries like Guatemala and Honduras, tends to see forests, flora and fauna as the victims of 'man's inhumanity to nature'. Humankind is considered the culprit and socio-economic activities need to be strictly controlled. Another approach which appears to characterize the activities of certain ecology groups that are particularly active in Nicaragua and El Salvador tends to see natural resource destruction in the context of 'man's inhumanity to man'. What is basically at fault is a specific social and economic system characterized by skewed patterns of resource distribution and exploitative social and external trade relations. Programmes to protect the environment, therefore, must be part and parcel of a broader development strategy which addresses more fundamental problems associated with poverty, inequality, the concentration of wealth and North–South relations.[1]

It would seem to be the case, however, that ground-level experience brings with it some degree of social awareness. Several groups which at the outset assumed a purely 'conservationist' stance have come to recognize that forest protection must go hand in hand with development projects and programmes that provide alternative livelihood opportunities for the rural poor. This transition has perhaps been most apparent in Costa Rica where, through time, a number of environmental organizations have adopted a more balanced approach combining conservationist with social or development concerns.[2]

While it is apparent that the conservation drive has gathered considerable momentum in recent years, there is still a gulf between government rhetoric and policy objectives, on the one hand, and the reality of policy and project implementation on the other. National institutions with responsibility for encouraging tree planting and forest protection still tend to be fairly weak. A major driving force behind many environmental protection initiatives are the pressures and inducements associated with certain types of external aid. Many positive initiatives in the field of environmental protection result from the fact that the temporary occupant of a key government post (or the spouse of a president) happens to be concerned about environmental issues and brings a green tinge to government policies and programmes.

Numerous environmental regulations and laws remain unenforced, conservation programmes unimplemented, while policies and projects intended to protect or rehabilitate the environment often impinge negatively on people's livelihood at the local level. There has been little assessment of the social impact of government or agency schemes which alter agricultural land-use patterns by encouraging producers to plant trees, that seek to promote sustainable forms of forest exploitation, or attempt to take large tracts of land out of the traditional arena of human exploitation by establishing protected areas.

To examine the social impact of forest protection and tree planting initiatives, local-level case studies were conducted in Costa Rica, Guatemala,

Honduras and Nicaragua. The existing literature on the problems encountered when implementing forest protection projects and programmes has tended to focus on a variety of technical, administrative or financial constraints associated, for example, with inappropriate project design and extension methodologies; limited budgets or foreign aid; and restrictions on qualified human resources.

While such aspects are referred to here and are indeed highly relevant, the central purpose of Part III of this book is to highlight the types of social and political economy questions which have undermined attempts to conserve forests and promote tree planting. What follows is a critical assessment of what has happened in practice when such initiatives confront social reality at the local level and encounter the real world of people, interest groups and politics.

The types of experiences examined in this study can be classified in terms of four very different approaches to forest protection and tree planting. These I have labelled and defined as follows:

- the 'conservationist' approach, involving attempts to introduce sweeping changes in land use and human settlement patterns through the creation of protected wildland areas;
- the 'project' approach, involving the implementation of numerous, relatively small-scale, programmes and projects associated with reforestation, sustainable forest management, agroforestry and social forestry;
- the 'revolutionary' approach, specific to Nicaragua during the 1980s, involving profound structural changes that attempt to transform patterns of inequitable resource distribution and human settlement which underpin shifting agriculture and the colonization of agrarian frontier areas; and
- the 'grassroots' approach, involving the organized efforts of local groups directly affected by deforestation to defend forest resources and ensure that the benefits derived from the exploitation of such resources accrue to themselves and future generations.

Chapter 7

The conservationist approach: National parks and reserves

The number of wildland areas in Central America officially declared as national parks and reserves increased from just 30 in 1970 to more than 300 in 1987 (Morales and Cifuentes, 1989:13). This rate of expansion has accelerated in recent years and, while accurate data are difficult to come by, it would appear that over 8 million ha, or one-sixth of Central America's total land area, are now officially designated as protected areas. Table 7.1 presents data on protected areas by country. Estimates for El Salvador and Nicaragua include existing and proposed sites.

While much has been done on paper to protect forests, the possibilities of effectively implementing environmental legislation concerned with protected areas are generally quite limited. Referring to the situation of the late 1980s, Green points out:

Table 7.1 Protected areas in Central America, 1990

Country	Protected area (millions ha)	% National territory
Panama	1.4[1]	18
Costa Rica	1.5[2]	29
Nicaragua	1.7[3]*	14
Honduras	2.5[4]**	22
El Salvador	–[5]***	–
Guatemala	1.7[6]	16
Total	8.8	18

Sources: 1. INRENARE, 1991
2. MIRENEM, 1990
3. Personal communication with Juan José Camacho of IRENA
4. SECPLAN *et al.*, 1990
5. Godoy, 1991 (see Annex 1)
6. Author's estimation based on Godoy (ibid.)

* This area corresponds to 43 protected areas, 20 of which are proposed sites. Most of the latter were due to acquire legal status during the early 1990s.

** This area includes the Olancho forest reserve of approximately 1.5 million ha.

*** Actual and proposed sites amount to no more than 50,000 ha.

> ... the steady growth in the number of parks and reserves has created the illusion that stretches of forest have been rescued from the onslaught of shifting agriculture. But there is an enormous gulf between legislating reserve status and enforcing it. Honduras and Guatemala, for example, struggle to manage two national parks apiece, while their numerous other protected areas are by and large abandoned to their inexorable fate. Indeed, the bitter truth is that throughout the isthmus many parks and reserves exist on paper alone, and the assault on the forest is often as severe within the parks as without. (Green, 1990:123–4)

This chapter assesses what had been achieved in each of the six countries by 1990 and examines the types of general problems which have affected the implementation of protected area status.[1] The experiences of Guatemala and Costa Rica are analysed in more depth.

PROTECTED AREA SCHEMES

Panama

Of a total land area in Panama of 7.6 million ha, 18 per cent or nearly 1.4 million ha form part of the country's national parks and reserve system. The first national park was established in 1966. In 1980, Panama set up what was then Central America's largest single protected area: the Darién National Park, which extends along the country's south-eastern border with Colombia and covers 579,000 ha. In 1987, a ten-year plan was drawn up to develop the country's national parks and reserve system which was to include 41 protected areas. Of the 25 protected areas which existed in 1990, only one (Barra Colorado) had actually implemented a management plan.

There is, generally, a tremendous gulf between what is planned and what can be achieved on the ground. Efforts to protect certain areas often lack the backing of legislation and are seriously constrained by limited personnel and financial resources. Moreover, the geographical boundaries of many protected areas have not been clearly defined and local populations living in or near protected areas have not been incorporated as participants and beneficiaries in protected area schemes. Of the 21 protected areas which existed in 1989, for example, only 13 had regular personnel. Of the 84 people who worked in these areas less than 9 per cent were professionals. In 1990 just 60 forest rangers were employed to protect 1.4 million ha.[2]

Whereas deforestation, erosion and sedimentation pose a major threat to the future of the Panama canal, the government has been reluctant to raise taxes to pay for watershed protection. With each ship that passes through the canal's lock system, approximately 2 million litres of water are discharged to sea (Collins, 1990:105). The Canal Zone authority, however, is reluctant to pay for the water. A proposal drawn up in 1990 by the natural resources institute INRENARE to raise the canal toll by three cents per ton sought to secure $9 million for protection schemes in the Río Chagres watershed. This proposal, though, met with considerable opposition.

Outside funding for forest protection is expected to increase substantially during the first half of the 1990s. A donors' conference organized during early

1991 obtained support for six important projects. In March of that same year the MARENA project, designed to create a conservation foundation and improve the management of forest areas, national parks and the Canal Zone watershed, was submitted to USAID. This project was formally valued at $44.5 million. It is important to note, however, that the amount of money which would actually reach the ground and be used to finance concrete forest protection activities is only a small proportion of the total. Only $8.3 million, to be spent over several years, was earmarked for technical assistance, construction and equipment.

Nicaragua

Prior to the revolution of 1979, the Nicaraguan government established three national parks. Only in one, however, was anything done to enforce protected area status. Following the creation of the Natural Resources Institute (IRENA) in 1979 more significant actions were taken to identify, establish and manage protected areas. A study concluded in 1983 identified 35 areas covering 17,265 square km, or 13.3 per cent of the national territory, as potential protected areas.

The intensification of both the civil war and economic crisis following 1983 prevented IRENA from advancing on the forest protection front. Instead, the institution was weakened as budgetary constraints and risks to the physical safety of staff working in isolated areas forced many qualified employees to leave and prompted a rapid turnover of personnel. The capacity of IRENA to influence the policy process was further weakened during 1988 when, as part of a cost-cutting state rationalization programme, it lost its ministerial status and became a department within the Ministry of Agricultural Development and Agrarian Reform (MIDINRA). Following the change of government in 1990, IRENA regained its ministerial status but its activities were constrained by budget cuts associated with a stringent economic stabilization programme.

This has severely affected conservation programmes. For example, the Bosawas forest reserve in north-eastern Nicaragua formally covers more than 7,000 square km but due to budget constraints, IRENA has had to concentrate its activities in 1,000 square km.[3] Implementing protected area status in the reserve and wildlife areas which make up the Sí-a-Paz project in south-eastern Nicaragua has also proved extremely difficult due to financial and human resource constraints. While extending over 3,000 square km, this area had just four full-time inspectors.

During the early 1990s, a number of national and international NGOs and (mainly Scandinavian) bilateral aid agencies assumed a more prominent role in forest protection initiatives. The major constraints which currently affect efforts to establish and manage protected areas include severe limitations on human, material, technical and financial resources; little understanding, participation and support of local communities and authorities in the identification, creation and management of protected areas; the increased pressures on forest resources emanating from national and transnational logging enterprises, as well as the rural poor and returning refugee populations seeking to rapidly colonize forest areas; serious lack of inter-sectoral and inter-agency co-ordination regarding the use of specific geographical areas and natural resources; and limited enforcement of existing legislation governing forest protection.

Honduras

Like most other countries in the region, Honduras has created numerous protected areas. This process, which began in 1961 with the introduction of legislation to protect mangroves in the southern Pacific coastal area, accelerated during the early 1970s when large areas were declared forest reserves. The Olancho forest reserve, created at this time, covers some 1.5 million ha and, on paper at least, is Central America's largest protected area. In 1980, Honduras established what was then the region's largest biosphere reserve, the Río Platano Biosphere Reserve, covering an area of 525,000 ha. One decree alone, issued in 1987, created 37 protected areas which included 11 national parks, 8 wildlife refuges and 18 biological reserves located in cloud forest areas. By 1990, 51 protected areas had been created and another 41 proposed. The total area which had acquired protected area status amounted to 2.5 million ha or 22 per cent of the national territory (SECPLAN *et al.*, 1990:207).

In practice, however, it has been virtually impossible to enforce protected area status in Honduras. The problem stems in part from the manner in which protected areas were created. As one official report observes:

> The majority of the 51 protected areas acquired legal status in an *ad hoc* manner, almost always without the prior elaboration of natural resource studies and often without having taken into account priorities based on technical criteria. The motives underlying the creation of the majority of these areas generally related to political considerations and in some cases personal interests. (ibid.:198 – author's translation)

Numerous legal, administrative and social problems affect the implementation of legislation governing protected areas. Several areas which have been declared as parks or reserves lack an adequate legal framework for actually enforcing protective status. Certain laws and regulations defining land-use patterns and the role of particular government agencies in a specific area are contradicted by others which define alternative uses and responsibilities. The majority of the laws which have created protected areas have not, as yet, been reinforced by regulations which are necessary for the law to be implemented.

While over 50 protected areas already exist, only five have drawn up management plans. In most of these areas, however, effective protection has not been achieved. The Río Platano Reserve, for example, is administered by just 12 people. Little has been done to integrate the numerous peasant and tribal communities which live in the area in a more sustainable development process. Another protected area which formally has a management plan is the Yojoa watershed. Here, however, the operations of different state institutions have contradicted one another. The National Agrarian Institute (INA), for example, has established peasant settlements in ecologically inappropriate areas.

El Salvador

As indicated in Part I, there are few forests left in El Salvador to protect. While the 1973 forestry law established the need to create national parks and reserves, it is estimated that just 7,000 ha are effectively protected. The National Parks Service, set up in 1981, administered only six areas in the late 1980s. Studies conducted during the 1980s identified 60 areas, covering 50,000 ha, which could potentially be integrated in a 'System of Natural Protected Areas'.

In practice, however, most protected areas have been subject to intense social pressures. The Montecristo National Park, for example, resembles more an integrated rural development project rather than a protected area. Several protected areas were rapidly deforested during the 1980s.

There are specific factors that have impeded the establishment of a system of national parks and reserves. These include restrictions on movement of public sector employees imposed by the civil war; the high cost of land; the social pressures for land redistribution; minimal national budgetary allocations or external aid for conservation; and the lack of participation of both peasants and the private enterprise sector in the environmental planning and policy process.

Guatemala

Initiatives in Guatemala to protect forest resources began as early as 1945. With the enactment of the first forestry law during the 1950s, 14 sites were declared protected areas. Measures to protect forest areas intensified during the 1970s with the introduction of a more comprehensive forestry law in 1974, the creation of the National Forestry Institute (INAFOR), and the emergence of several conservation organizations concerned with the rapid rate of environmental degradation in Guatemala. The University of San Carlos assumed an active role in administering several biological reserves or 'biotopes'.

When civilian government was formally restored to Guatemala in 1985 some advances were made in the area of social and environmental policy. In 1987, the Centre for Conservation Studies (CECON) of San Carlos University, with support from the World Wide Fund for Nature (WWF), established three more biotope reserves in the Petén region. Until the 1990s, however, protected area sites were generally quite small: by mid-1989, for example, some 52 protected areas existed but they covered only 2.7 per cent of the national territory.

During 1989 and 1990, the conservationist drive in Guatemala took a quantitative leap forward. The Protected Areas Law of 1989 laid down more rigorous guidelines for establishing protected areas and established the National Council of Protected Areas (CONAP) to oversee the country's protected areas system (SIGAT). In 1990, two large biosphere reserves were created in the Petén (1.4 million ha) and the Sierra de las Minas (236,000 ha).

With these developments, some 16 per cent of national territory acquired protected area status.

A new municipal code passed by the Guatemalan Congress in October 1989 identified specific roles for municipal governments in the area of environmental protection. It remains to be seen, however, whether protected area status can be effectively implemented. Economic and political power remain highly concentrated in the hands of large landowners, a few industrialists and the military. Such groups constitute powerful interest groups which can effectively undermine progressive legislation and government policies (Bradley *et al.*, 1990b). Thus, while several important initiatives were taken by the Cerezo administration in the area of environmental protection, there is considerable doubt regarding the state's capacity to implement relevant legislation and policies. A USAID report on natural resource policy in Guatemala summed up the problem in the following terms:

> The country's legal base, institutional framework and citizen participation in the policy process are too weak to effectively implement policies related to environmental issues.... Laws are often outdated ... fines for violations are too low, license systems arcane, environmental impact assessments are not required, educational programs are not included, and mechanisms for enforcement of the law nonexistent.... Institutions mandated to implement policies lack both the credibility and the long-term financing necessary to be effective.... Instead of an integrated and coordinated legal system to protect the environment, each institution involved ... issues its own individual and often outdated policies.... Centralized decision-making, erratic actions, and biased political agendas would make any rural population highly skeptical of any official central government programs. However, the case in Guatemala is acute. The scepticism of the Indians, generated from a long history of centralized decision-making dominated by *ladinos*, broken political promises, and outright exploitation and violations of human rights has translated into apathy and non-participation in the political arena. (Bradley *et al.*, 1990b:11, 23–4)

The case of the Petén

The difficulties in both designing and implementing legislation regarding protected areas is brought out clearly in the case of the Petén in northern Guatemala. This region covers one-third of the national territory and has half the country's forest reserves. Historically, it has remained marginal to the process of national development and isolated from the rest of the country.

While the total population of the area numbers only a quarter of a million, the biosphere project has had to confront a complex social reality. A variety of social groups and economic agents have interests in the area. For decades the area was administered by the military which distributed territory and exploited the region's natural resources largely for the personal gain of military officers. It was not until the late 1980s that civilian authorities began to administer the area.

While the area lacked basic economic and social infrastructure and was not suitable for basic grain production, it came to be regarded nationally as a sort of 'promised land' due to the fact that unclaimed land was plentiful. The population of the Petén has increased tenfold since 1964 to approximately 250,000 (Plan de Acción Forestal Tropical para Centroamérica, 1991:64).

Cattle ranching, on large estates, expanded rapidly in the southern part of the Petén from the 1960s onwards and was a major factor contributing to the country having one of the highest rates of deforestation in Central America.

Throughout this century, national and foreign lumber companies have extracted precious hardwoods such as mahogany and cedar. During the 1980s, nine lumber companies operated in the area, systematically ignoring legislation and regulations governing the use of forest resources. For decades, the oil industry, dominated by United States companies, has also operated in the area.

The tourist industry, with its central focus on the famous Mayan site of Tikal, has significant interests in the region. Moreover, there is an important clandestine economy centred on the theft of relics from the regions numerous Mayan archaeological sites, contraband in wood to Mexico, consumer goods from Mexico and Belize, and live animals to Guatemala City and abroad.

Three other activities have provided employment for thousands of families. An estimated 7,000 people cut *xate*, a small ornamental plant which is exported to the United States and Europe. Sales of *xate* were valued at approximately $2 million in 1987. In northern Petén, local people also produce *chicle* or gum for export, mainly to Japan. Black pepper (*pimienta gorda*) gathering also constitutes a source of income for some families.

From the 1960s to the mid-1980s, the natural resource base of the Petén was rapidly degraded. This situation began to change in the late 1980s. In 1988, DIGEBOS assumed responsibility for administering the region's forests and in 1989 CONAP was created to set up and manage the system of protected areas. Numerous national and international conservation organizations flooded into the area to conduct studies and provide different forms of support for protected areas. By 1991, some 13 units made up the Protected Areas System of the Petén (SIAP) while another 21 areas had been identified as potential protected areas. The biosphere reserve contained a nucleus area of 800,000 ha and 'multiple-use zones' of 650,000 ha. Of the sites which make up the SIAP, only five, however, had implemented a comprehensive management plan. In the remaining areas little or nothing has been done. While dozens of agencies and millions of dollars in aid are now pouring into the Petén with the objective of protecting the region's flora and fauna, major problems threaten the implementation of the project, and even its desirability has come into question.

First, the capacity of the state to regulate and co-ordinate the activities of all these agencies is extremely weak, due not only to the usual resource constraints which characterize state institutions in the country, but also to the fact that central government agencies have only recently begun to operate in the region. Moreover, the actions of various state agencies and the orientation of certain economic policies are inherently contradictory. The conservationist approach confronts a style of economic development and a set of public policies which underpin the colonization of agrarian frontier areas, cattle ranching and the production of basic grains. They also encourage activities such as tourism and the 'mining' of the region's natural resources which are organized in such a way that few benefits remain in the region.

Second, there has been a tendency for certain government agencies and national and foreign NGOs to ride roughshod over local authorities and social groups. Conservation projects and programmes often fail to consider the interests of local residents both at the planning and execution stages. During early 1991, for example, local municipal authorities opposed the biosphere's largest project, the $22 million USAID-supported MAYAREMA project. They believed the project would undermine their authority by strengthening the hand of CONAP. These authorities also had the support of local ranchers and other groups who saw the project as a threat to their livelihood.

Third, the conservationist approach has to confront the reality of poverty and the fact that thousands of families living in the area have no alternative but to exploit the region's natural resources. Moreover, an estimated 250 people from other areas of the country arrive daily seeking land and employment in the region.

Fourth, the armed conflict which has affected certain other areas of the country has extended to the Petén. Military repression has served to further alienate local populations and increase mistrust *vis-à-vis* state institutions. The operations of both the guerrillas and the army between 1988 and 1991 affected the implementation of several projects in the region.

Finally, the clash of different cultures could not be greater – what might be called a case of conservationists or 'greens' versus *ladino* 'cowboys', with a small Indian population to boot. The Petén was pioneer territory par excellence with all that the term implies in terms of a 'get rich quick' mentality, lawlessness and corruption. To transform such a world clearly requires a more integral approach than one which emphasizes strict control of the use of natural resources and the exclusion of people from protected areas.

One of the most surprising aspects of the conservationist approach not only in Guatemala but throughout most of the region is that complex social realities such as these are largely ignored in the studies which promote the creation of protected areas. It is often assumed that dealing with local residents will be a straightforward matter and that conservation will result in tangible benefits for the local economy and society. The possibility that social tensions and conflict might arise is rarely considered.

This is apparent in the technical study which justified the creation of Guatemala's second largest biosphere reserve, the Sierra de las Minas, which was established by law in 1990. Unlike many other such studies, there is some discussion of the complex socio-economic situation of the area: Kekchi Indians who live along the reserve's northern border; *ladino* peasants and larger commercial farmers in the south; and several lumber companies with sawmills dotted throughout the zone. Yet there is no reference to any potential problem associated with regulating their activities and rapidly transforming traditional economic and cultural practices (Fundación Defensores de la Naturaleza, 1989).

Costa Rica[4]

Since the creation of the country's first reserve in 1963 and the enactment of the 1969 Forest Law, more has been done in Costa Rica to protect forest resources than in any other Central American country. Eighty national parks and reserves of different categories, covering 1.5 million ha, or 29.3 per cent of the national territory, were legally constituted as protected areas by 1990 (MIRENEM, 1990:19–20). Some 65 per cent of these protected areas is covered with forest.

In addition to protecting forest resources, these areas have been created with the objective of preserving biodiversity and natural habitats (national parks, biological reserves and wildlife refuges); protecting the environment through the sustainable use of natural resources (forest reserves and protectorate zones); and protecting cultural forms, life systems and methods of resource use in areas where Indian populations are concentrated (indigenous reserves).

NGOs have played a major role in the establishment and administration of protected areas and in supporting numerous projects in these areas. The role of foreign NGOs has been reinforced by the fact that several international agencies have their Central American headquarters in Costa Rica. Political stability and democracy have given Costa Rica a comparative advantage in obtaining external aid and support from many international agencies. The country's forest protection programmes also enjoy considerable public support, due partly to a relatively effective programme of environmental education over the past two decades.

An important development that has strengthened the capacity of the state to undertake forest protection and environmental protection initiatives during the 1980s has been the process of institutional consolidation. It was noted earlier that the existence of weak institutions has been a major constraint which has undermined the conservation approach in the different Central American countries. The situation in Costa Rica is somewhat dif-

Table 7.2 Protected areas in Costa Rica, 1990

Category	Total area (ha)	% of national territory
National parks	404,906	7.9
Biological reserves	17,514	0.3
Wildlife refuges	128,027	2.5
Forest reserves	336,890	6.6
Protectorate zones	126,200	2.5
Frontier zones	120,275	2.4
Mangroves	35,000	0.7
Indigenous reserves	326,483	6.4
Total	1,495,295	29.3

Source: MIRENEM, 1990:20.

ferent. The main agencies with responsibility for protected areas (other than the indigenous reserves) – namely the National Parks Service (SPN), the General Forestry Directorate (DGF), and the Wildlife Division – moved from the Ministry of Agriculture and Livestock (MAG) to strengthen the Ministry of Natural Resources, Energy and Mines (MIRENEM) which was set up in 1988.

Moreover, several agencies with responsibility for protected areas have acquired greater authority and status. Such was the case of the National Parks Service which is now a directorate within MIRENEM but which, prior to 1977, was merely a department of the Forestry Division. As a directorate, the SPN acquired its own budget and became less dependent on extraordinary forms of support such as the backing of the First Lady during the Figueres presidency (1970–74). By the late 1980s, the SPN had become the most important government agency with responsibility for forest protection and conservation. The SPN, however, has received relatively little government funding. During the late 1980s, central government budgetary allocations amounted to approximately $1 million, the bulk of which went to pay salaries. The agency, however, was able to augment its administrative capacity by mobilizing human and financial resources from a variety of national and international sources. Particularly important in this respect was the creation of the National Parks Foundation (FPN), set up in 1980 to attract and administer funds that were flowing in from international, mainly United States-based, conservation organizations (Boza, 1987).

During the late 1980s, the existence of the FPN facilitated the reception of funds via the debt-for-nature swap mechanism. The latter involves the conversion of foreign debt titles, available at a discount on the secondary market, into local currency for investment in conservation programmes (Brüggemann, 1990).[5]

This mechanism was instituted in 1987 with the purpose of reducing the national debt, generating funds for environmental protection, and, through eco-tourism, involving the business sector in conservation. The debt-for-nature swap mechanism enables conservation agencies (eg WWF, Nature Conservancy) or bilateral agencies (eg Sweden, the Netherlands) to purchase foreign loans at a discount, on the understanding that the national government will provide funds in local currency at a dollar equivalent which exceeds considerably the cost to the purchaser. Such funds are used to purchase national park and reserve land, provide financial incentives for reforestation schemes, as well as institutional support for environmental organizations (*La República*, 4 August 1990). By mid-1991, Costa Rica accounted for 44 per cent of the total value of the world's debt-for-nature swaps. Between 1987 and mid-1991, some $80 million of national debt was purchased for $12.5 million and converted into $42 million of local currency commitments (see Table 7.3).

By the early 1990s, the initial enthusiasm for debt-for-nature swaps had waned to some extent. What, in World Bank parlance, had seemed like a definite 'win–win' situation – in which all those involved (international banks, the national government, NGOs, environmentalists, local landowners)

Table 7.3 Costa Rica: Debt-for-nature swaps, 1988–mid-1991

Date	Purchaser	Cost to purchaser ($)	Face value of debt ($)	Funds generated $ equiv.[1]
1988	Netherlands	5,000,000	33,000,000	9,000,000
1988	FPN/WWF	918,000	5,400,000	4,050,000
1989	Sweden	3,500,000	24,500,000	17,100,000
1989	NC	784,000	5,600,000	1,680,000
1990	Sweden/WWF/NC	1,953,473	10,753,631	9,602,904
1991	RA/MCL/NC[2]	360,000	600,000	540,000
Sub-total		12,515,473	79,853,631	41,972,904
World total		16,705,000	98,445,000	61,064,000

Source: Based on Mahony, 1992, Table 1, p. 98.

1. Dollar equivalent of local currency. Conservation funds generated do not include interest on bonds when government pays in bonds rather than cash.
2. Debt donated by Bank of America.

FPN: National Parks Foundation of Costa Rica
WWF: World Wide Fund for Nature
NC: Nature Conservancy
RA: Rainforest Alliance
MCL: Monteverde Conservation League

stood to benefit – turned out, in practice, to involve a number of trade-offs, unexpected costs and a somewhat skewed distribution of benefits.

Clear winners in this exercise have been the northern financial institutions. As Mahony points out, not only have they been able to recover some of the debt, but they have done so by employing mechanisms (including not only debt swaps but also debt restructuring) which ensure that the discount rate on outstanding debt rises as the debt is paid off. As the price of the debt on the resale market increases the government is likely to find itself in a situation of having to repay the same amount as before the debt-for-nature swap took place (Mahony, 1992:99). Hence the benefits to the Costa Rican government are not as straightforward as one might assume.

In the case of Costa Rica, the value at which one dollar's-worth of debt was purchased on the secondary market increased sharply from 13 cents in January 1989 to 51 cents in November 1991 (CEPAL, 1991, Table 20). Under such conditions it is to be expected that the debt-for-nature swap phenomenon will be relatively short-lived. This is because the increase in the discount rate will scare off potential purchasers, and the government will be reluctant to comply with the so-called multiplier effect whereby it commits considerably more than the sum paid by the purchaser.

In actual fact, the amount of debt which has been cancelled through the debt-for-nature swap mechanism is only a tiny fraction of the country's total debt. In 1990, Costa Rica's total external debt amounted to $3,772 million

(World Bank, 1992). As indicated in Table 7.3, debt-for-nature swaps accounted for just $79 million or 2 per cent of the total debt.

By making available large amounts of local currency, debt-for-nature swaps also run the risk of fuelling inflationary pressures. Clear tensions have arisen in the context of economic stabilization programmes which attempt to control the money supply and reduce government deficits. In 1990, a debt-for-nature swap programme was approved which would have converted 100 million dollars of debt over a five-year period. Later that year, however, the programme was temporarily suspended and scaled down by the Central Bank given the fear that large-scale purchases of land would have inflationary effects (Bradley *et al.*, 1990a, Vol. 2:102).

Doubts have also arisen concerning the ability of the government to honour its commitments and the implications of debt swaps for other forms of government spending on environment. Donations are often in the form of bonds which are redeemable in several years' time. As Mahony points out:

> There is no guarantee that Costa Rica will honour these bonds . . . any more than it will pay back the rest of its debt. . . . But even if the government did honour this new debt, it could cut its other environmental spending to make up the cost. (Mahony, 1992:100)

Perhaps the major question mark surrounding the debt-for-nature swap mechanism concerns the capacity of government agencies and NGOs to effectively implement protected area status and to do so in a way which benefits local populations. It is this aspect which is considered in more detail in the following chapter.

While Costa Rica is often held up as a sort of model for other countries in the region, forest protection initiatives have experienced serious limitations. While the conservation drive began in earnest during the late 1960s it proved extremely difficult to contain deforestation over the following two decades. Most of Costa Rica's forests are now located in protected areas but the expropriation of private holdings in such areas still has a long way to go. According to the Costa Rican Tropical Forestry Action Plan: 'real protection in many of these areas is relatively weak' since current regulations still permit landowners to exploit forest resources (MIRENEM, 1990:5).

Since the 1960s, numerous laws have been introduced in Costa Rica to protect forests. A study conducted by the Costa Rican environmental group CEDARENA has shown, however, the limited effectiveness of much of this legislation. While legislation passed in 1961 (Ley de Tierras y Colonización), for example, attempted to bring a halt to spontaneous settlement of public lands:

> this provision in the law has been largely ignored by settlers and weakly enforced by the government. . . . The history of toleration of mass invasions of public lands in the three decades since the public domain was placed legally off limits to settlers . . . suggests that the state is not serious about excluding settlers from state land. (CEDARENA, 1990)

The legislation and procedures governing the establishment and administration of protected areas is flawed in numerous respects. While the process of creating such areas requires a preliminary study of the demographic and land tenure situation in the area, it does not automatically involve a socio-economic study of the population which will be affected. Once a protected area is established, private land within the boundary of the area must be purchased. This procedure demands enormous resources. The state must compensate not only those with legal title to the land, but also those with possession rights or who have made improvements on the land (fences, planting of perennial crops, etc.). About 10 per cent of the area administered by the SPN is still privately owned (DEA, 1990) while the percentage is even higher in most other categories of protected areas.

The legislation ignores the fate of those who must leave the protected areas. Once paid, residents must fend for themselves elsewhere. Only in a few instances has the state facilitated access to land elsewhere or alternative employment.

The enforcement of regulations governing land use in protected areas has been generally weak. The penalties for infringements of the laws vary from fines to prison sentences ranging from six months to three years in the case of forest fires. However, both the lack of control and the nature of the legal system often mean that penalties do not act as a deterrent. The legal system is both extremely slow and, at the local level, culprits are often treated lightly. The lack of control is a function not only of limited human and material resources such as vehicles, but also the policy of forest protection authorities to deflate social tensions. Controlling deforestation in the indigenous reserves has proved difficult. The forestry department has no inspectors in these areas. According to the Indigenous Law, land in the reserves is common property, to be administered by local community organizations known as Integral Development Associations (ADIs). The ADIs must authorize the felling of trees. In practice, however, many people fail to obtain the required authorization.

Not only have state institutions and NGOs been severely stretched in their efforts to enforce protected area status but, for many years, their activities remained uncoordinated. It was not until the late 1980s that the state attempted to devise a more coordinated and systematic approach by creating Regional Conservation Units. These are bodies which oversee the work of the different institutions and agencies working in a group of protected areas in close proximity to one another. By 1991, 11 such units were in existence.

Chapter 8

Protected area schemes and social conflict in Costa Rica[1]

Probably the most complex aspect which undermines the conservationist approach in Costa Rica relates to tensions and conflicts which arise when the interests of different groups or agents are affected by regulations governing protected areas. Social conflict in and around protected areas has generally been associated with four aspects: the way protected areas have been established; the process of land acquisition; the invasion and occupation of land in protected areas; and resource use in such areas.

It has been usual for protected area status to be imposed on a specific area without prior consultation with the local population. The Barra del Colorado Wildlife Refuge in north-eastern Costa Rica, for example, was created in 1985 without the knowledge of many people living in the area. Many local residents were somewhat surprised when inspectors arrived and announced that traditional activities such as hunting and tree cutting constituted illegal activities and that the state intended to expel those who did not possess land rights. The local population, however, became organized and halted, for some time, the enforcement of these regulations.

Lack of adequate consultation and clarification concerning the demarcation of protected area sites, as well as the limited capacity of conservation agencies to enforce protected area status, particularly during the initial phase of a protected area project, can lead to accelerated deforestation. A 'get what you can, while it lasts' mentality has been evident in certain areas. Referring to the experience of protected area schemes in north-eastern Costa Rica, Brüggemann writes:

> Moves by the state and international conservation organizations to establish a bi-national park with Nicaragua and create a biosphere reserve ... acted as an incentive [to many farmers] to cut and sell timber before it became impossible. (Brüggemann, 1992)

For three years the 92,000 ha Barra del Colorado wildlife refuge existed as a so-called 'paper park'. In 1988, however, stricter controls were implemented on agricultural expansion and tree felling, while squatters and more established settlers were no longer able to obtain land titles. Once the 1986 Forestry Law came into effect, it became extremely difficult for farmers to acquire the numerous certificates and documents (including a land title) required for obtaining a tree felling permit. Operating in the area, however, were several agents, including the large lumber company, Portico, and a number of independent timber traders or '*madereros*'. They had the necessary financial and administrative resources, as well as connections with local authorities, to acquire the documentation required by the Forestry Law for tree felling. They would sometimes buy out farmers both inside and outside the reserve area, and through various devices manage to obtain title to the land. Reports from the area suggest that tree felling operations often spilt over onto the land of other farmers who were eager to sell their trees, albeit at 'sub-market' prices, out of fear that government controls on land use would intensify in the future (Brüggemann and Salas, 1992).

The studies which justify the establishment of protected areas have tended to focus more on biophysical aspects and neglect others relating to patterns of land use and the social situation of populations living in and around the reserve. Several laws and decrees governing protected areas have had to be amended at later dates to accommodate the socio-economic reality of the area in question. The Arenal Forest Reserve decree, for example, was modified two months after being introduced to exclude land that was already in agricultural use (Kauck and Tosi, 1990). There is, however, increasing awareness on the part of NGOs and state institutions concerning the need to consider social aspects and to encourage the participation of local residents in the planning process associated with the establishment of protected areas.

Sometimes, local groups have organized to defend their interests when they have learnt that plans are afoot to create or expand a protected area. This occurred, for example, in the Tortuguero National Park in north-eastern Costa Rica. Moves to expand the area, by the IUCN and other international conservation NGOs in co-ordination with MIRENEM, prompted local groups, including small farmers and settlers, as well as the timber company Portico, to form an organization to defend their interests.

As indicated in Chapter 7, once an area has been declared a protected area, the state has the obligation to purchase private sector land. The process of indemnification is a major source of conflict. Shortages of funds induce the state to conduct tough negotiations and delay payments. When the property value cannot be mutually agreed upon, the property is formally expropriated and compensation is decided via a lengthy legal process that can take several years and terminate in payment by government bonds. Whether or not landowners and farmers get a fair deal will often depend on their capacity to organize and/or obtain legal assistance.

This is brought out clearly in the case of the expansion of the Santa Rosa national park and the creation of the Guanacaste national park during the late 1980s. The private individuals affected made up three groups: squatters who had acquired right of possession, large absentee landowners, and small and medium-sized landowners. The economically powerful group of larger landowners negotiated individually but with legal assistance. They were generally able to obtain payments in the region of 50,000 colones/ha (approximately $600). The group of squatters was well organized and negotiated as a group with the assistance of a lawyer. The FPN, which bought land on behalf of the state, apparently tried to weaken the group by threatening expropriation and manoeuvring a split on the issue of who should lead the group. Eventually a price of 14,000 colones/ha ($170) was agreed upon. This fell far short of the 50,000 colones/ha originally demanded by the group but exceeded the amount obtained by others.

The group of small and medium-sized landowners, consisting of farmers scattered throughout the area, was unable to organize itself effectively and obtained a low price for the land. This group also included squatters without any legal claim to the land. Those without property rights were paid for investments or 'improvements' they had made to the areas they farmed. Those with perennial crops, for example, received 2,000 colones/ha ($24).

Once property values are agreed upon, the purchasing authority often draws out the payment process and sometimes pays in bonds. In 1989, the government still owed landowners more than a billion colones ($11.8 million).

Squatters and landowners located within the boundaries of a protected area must resettle and either buy land elsewhere, invest in other economic activities or find work as wage labourers. Many choose to resettle in adjacent areas but it is often the case that agricultural land is extremely scarce or that land prices rise sharply once a protected area is established. Reforestation incentives have also forced up land prices in areas adjacent to parks and reserves. As such, many small farmers must resettle on even smaller, poorer quality, sites. This situation has forced producers in certain areas to invade protected areas in search of land.

The invasion of protected areas by land-seeking peasants, speculators and gold-miners is illegal and has provoked major conflicts both with the state and others, such as Indian populations, who have the right to live in protected areas. Land speculators may pose a particular problem during the initial phase of a protected area scheme. People may move into the area, stake a claim to the land, quickly convert forest to agricultural land and benefit from the compensation which is subsequently paid. This may be possible due both to the lack of control by the authorities and certain legislative loopholes.

In the case of the Arenal Forest Reserve, for example, people moved into the area after the reserve was formally established. They could obtain legal claim to the land because the law accepts the testimony of neighbours 'proving' that someone has resided in the area for at least 10 years prior to the formation of the reserve. The lack of aerial photographs and survey data make it difficult for the forest protection authorities to disprove fraudulent claims (Kauck and Tosi, 1990). Some of those who moved in were peasants who, having been displaced by a hydroelectric scheme in the area, were in dire need of land for subsistence production. Others, however, were dairy farmers located in areas adjacent to the reserve. They invaded reserve areas and cleared forest land in order to extend their pasture area. While the capacity of the DGF to control situations such as these improved through time, it nevertheless found it difficult to enforce regulations, given the fact that those occupying lands in the Arenal reserve organized with the national union of small farmers (UPA Nacional) which supported their cause.

Another group affected by the creation or expansion of protected areas consists of labourers who lose income-earning opportunities without indemnification. A few local employees may find work with the new national park or reserve authorities but most have to migrate to find alternative employment. In the case of the large Guanacaste national park, many workers migrated to the capital San José and other urban centres in Costa Rica's Central Valley region.

The social situation is particularly complex in the Indigenous Reserves where much of the Indian population has yet to perceive the benefits which derive from forest protection policies and schemes. Tensions often develop not only between the state and Indian populations but between the latter and 'whites' who have settled in the reserves. Significant areas in the indigenous reserves are occupied by settlers who purchased land or moved there before the reserves were created. An estimated 34 per cent of the land designated as indigenous reserves is occupied by settlers. While most of these people purchased land from indigenous people before the declaration of reserve status, some illegal land transactions have occurred following the establishment of the reserves.

Those who have invaded the indigenous reserves of La Amistad region have included peasants who have been displaced from their land through the creation of protected areas elsewhere (as in the case of the Chirripo National Park) or by people employed, or formerly employed, on the pineapple plantations of the United Fruit Company in Pacific coastal areas just outside the reserve area.

As indicated in Chapter 4, the Costa Rican state has not taken decisive action to defend the rights of Indian groups. This is reflected in the advance of settlers on Indian lands and the active role of lumber and mining companies in several reserve areas, not to mention the extent of poverty and the sharp disparity in the living levels of Indians and other social groups in Costa Rican society.

Restrictions on the use of natural resources in protected areas constitute another source of social conflict. Not all groups have equal access to the incentives which are offered to those who submit land to the forest regime. In the case of the Barra del Colorado Wildlife Refuge, for example, it is apparent that incentives have favoured mainly larger landowners. Small farmers lacking property titles have been unable to obtain various incentives, such as subsidized credits. Moreover, the Peasant Forestry Development programme which provides incentives to small farmers for reforestation does not operate within the reserves.

Apart from lumber companies, farmers and Indians who continue to use natural resources in protected areas, the main economic agents with interests in such areas are generally those associated with hydroelectric schemes and mining. Such operations have led to serious disputes with other interest groups. The proposed Río Pacuare hydroelectric scheme in one of the country's protected areas has provoked a conflict between the state-owned electricity institute (ICE) and an alliance of conservationists, tourist agencies which use the river for white-water rafting tours, and local indigenous families who would have to be resettled if the project goes ahead. When the institute initiated geological studies in the area, these groups formed a defence committee and organized an on-site protest in 1990 to block the project. The protest received considerable media coverage and forced a temporary suspension of ICE's activities in the area. A month later, ICE reached an agreement with the committee whereby the former could continue to conduct studies in the area while a commission would be established to attempt to harmonize the interests of the different parties involved (Dudenhoefer, 1990).

A similar dispute arose in 1989 when the transnational mining company EUROSPECT obtained a concession to mine sulphur in 400 ha of the Juan Castro Blanco protectorate zone. Before commencing mining operations, however, the company had to conduct an environmental impact study that had to be approved by a government commission. A local conservation organization was formed by residents of a nearby town to defend the area. This NGO worked both at the technical level, obtaining scientific information regarding the environmental impact, and at the political level, gathering support from local residents, as well as politicians and conservation organizations. The environmental impact study was subsequently rejected by the government which several months later also cancelled the company's concession in the area. The local NGO is currently attempting to force the government to declare the area a national park, an action which would deny mining companies the right to apply for concessions (CENAP, 1990).

TWO CASE STUDIES: THE CARARA BIOLOGICAL RESERVE AND THE OSA PENINSULA

To examine in more detail the nature of social conflicts which arise when conservation areas are created or expanded, case studies were undertaken of two zones: the Carara Biological Reserve in the central-Pacific region of Costa Rica and the protected areas on the Osa Peninsula in the south-western tip of the country (see Map i.1).

The Carara Reserve

While the Carara Reserve cannot be considered one of the hot spots of social conflict in Costa Rica, several problems have arisen which reflect some of the fundamental weaknesses in the conservationist approach which has characterized Costa Rica's environmental protection policy during the past two decades.

The reserve was created in 1978 in an area of 7,700 ha. Its aim was to protect the last remnants of forest and the considerable biodiversity which exists in what constitutes a transition zone between the drier northern region of the country and the more humid region to the south.

Local residents had always used the area for hunting and, to a lesser degree, for forest products. Tensions soon arose when the authorities attempted to enforce regulations. During the early years, forest guards adopted an aggressive attitude which aggravated frictions between the communities and the reserve administration. Local residents recalled how hunting dogs were occasionally shot by guards and how the refrigerators of local restaurants or *cantinas* were searched for illicit supplies of meat.

There was no attempt to encourage the participation of the community in conservation activities. It was not until the mid-1980s, when a Peace Corps volunteer began an environmental education campaign, that someone thought to explain the *raison d'être* of the reserve to the local population.

Enforcing regulations, however, proved difficult. Whereas the reserve management plan estimated that approximately 30 guards would be required, only seven existed. While the hunting activities of local residents declined over time, professional poachers from the Central Valley increased their activities as demand for live exotic animals grew.

The reserve contained both dense forest and two degraded areas in the mountainous southern and eastern areas of the reserve. Peasant farmers produced basic grains in the southern part while land-seeking peasants subsequently invaded the eastern part. When the reserve was created, little account was taken of the situation of those who used the land. Peasants in the southern area struggled for six years to force the Institute for Agrarian Development (IDA) and the SPN to establish settlements. Peasants in the eastern zone were able to enlist the support of a politician to back their cause. In 1983, nearly 3,000 ha of the reserve were handed over to the IDA for settlements.

The creation of these settlements clearly reduced pressure on the forest area but did little to improve the living conditions of the peasants involved. The settlements received some technical, material and financial support from government agencies such as the IDA and Ministry of Agriculture (MAG) as well as the EEC which implemented a rural development project. Much of this support, however, was highly inappropriate for the types of producers and bio-physical conditions of the area. Subsistence peasant producers with poor education and using rudimentary technology were suddenly expected to engage in efficient cash crop production of fruit and vegetables, employing sophisticated modern inputs. Goats and bees were introduced: the goats ate the crops and, according to some accounts, the bees left the area in search of better conditions. Moreover, no account was taken of the buffer zone character of the area; the need, for example, to grow crops which would not be eaten by natural predators from the forest was overlooked, as was the appropriateness of reforestation and agroforestry schemes.

Conflicts of interest also developed between the SPN which administers the reserve and the EEC project supporting a co-operative settlement located to the north of the reserve. In order to improve access to the settlement, the agencies supporting the settlement pushed for a road to be built through part of the reserve in order to avoid the more costly alternative of constructing a bridge over a nearby river. While an environmental impact study was conducted to determine what impact the road might have, no one foresaw the growth in the number of tourists visiting the area. Visits to the reserve rose from 5,000 in 1987 to 15,000 in 1990, creating a situation which the reserve authorities could not control. Not only has wildlife in the area been affected by the influx of tourists, but local communities have not benefited from the tourist boom. There has been no attempt to involve local residents in tourist projects. By law, it is illegal for local residents to engage in commercial activities in national parks or reserves.

The Osa Peninsula

The Osa Peninsula has been the site of intense conflicts arising from the establishment of conservation areas. The area is extremely rich in natural resources with a tremendous variety of flora and fauna, and relatively large expanses of dense forest. It is also rich in one other resource which has lured thousands of people to the area – gold. The peninsula is the site of two conservation areas totalling approximately 110,000 ha. These are Corcovado National Park, created in 1975, and the Golfo Dulce Forest Reserve, established in 1978.

The forest reserve has been subject to pressures from those seeking land. The construction of a road to the peninsula and the withdrawal of the United Fruit Company from a nearby area increased the flow of peasants and former plantation workers to the reserve in search of land and income from the sale of logs to the local timber industry. Social conflict has been minimized as a result of the limited capacity of the reserve authorities to enforce regulations.

Other problems in the forest reserve have arisen with peasant farmers in an IDA-settlement located in an area that subsequently became part of the reserve. When this land acquired reserve status, no property titles could be issued. This in turn made it extremely difficult for the farmers to obtain credit. Such a situation can have a number of negative side-effects from the point of view of conservation. In certain other reserve areas, for example, banks have begun accepting cows, rather than property titles, as collateral. This can encourage producers to expand pasture areas. When access to credit is blocked, it might also be expected that some producers who need cash at the beginning of the growing season will turn to illegal logging as an alternative source of finance.

The main conflict in the Osa Peninsula centres on the situation of gold miners in and around Corcovado. When the park was created, gold miners in the area were obliged to leave. They generally settled on the eastern border of the park. When the park was enlarged in 1980, they were forced to leave again but repeated attempts to expel them failed – they simply returned to resume their activities within the park area.

It was not until 1986, 11 years after the creation of the park, that an agreement was reached with the miners. In return for leaving the area, they were offered one of three alternatives: assistance in the formation of gold mining co-operatives outside the park but inside the forest reserve; plots of agricultural land in IDA-settlements; or a cash payment of 250,000 colones (approximately $5,000). As might be expected, the majority chose the cash payment.

All three alternatives failed: the co-operatives were generally unprofitable; miners with little agricultural experience soon chose to sell their plots of land; while many of those who had opted for the cash settlement failed to use it productively (reports from the area indicate an increase in the incidence of alcohol consumption and prostitution).

Two years later, the gold miners once again invaded the park. This time the SPN adopted a more repressive stance, and with the backing of the police, used force to expel them. Only a few miners, however, were actually prosecuted and many returned soon after being expelled. Since 1988, conflicts between park guards and the miners have intensified. In September 1989, two guards were kidnapped following the shooting of a miner.

These confrontations prompted MIRENEM and SPN to search for a more durable solution to the problem. Attention was focused, for example, on the design and implementation of productive projects associated with sustainable agriculture and timber processing. Also considered was the possibility of cancelling the concessions of certain transnational corporations, operating in nearby areas, to enable the miners to pursue their activities.

The case of Corcovado highlighted the limitations of the 'kick them out' approach to enforcing protected area status. The gold miners returned repeatedly, social conflict escalated, and nearly a decade later the government was forced to consider a very different approach.

The Costa Rican study also revealed the limitations of the 'buy them out' approach both from the point of view of social groups that are economically and politically weak and in the case of countries experiencing high inflation and pressures from international agencies to impose tight monetary policies.

What emerges clearly from the above review is that the conservationist approach has failed to come to grips with crucial social issues. It is common for somewhat technocratic formulas to be imposed which generally ignore the situation of thousands of families who live in the areas affected, depending on forest resources for subsistence provisioning and income. Protected areas have often been established without the participation, or even the knowledge, of local communities. Social impact studies, to examine how the creation of a conservation area will affect the local population, are rarely conducted.

Establishing protected areas is an extremely complex undertaking not only because it involves fundamental changes in land-use practices and human settlement patterns but also because it involves extending the arm of the state and development agencies into areas where suspicion of such agents, or overt opposition, is likely to characterize the attitudes of local inhabitants. People in outlying forest areas have often experienced a history of isolation and marginality, if not repression. Social groups living in such areas are probably amongst those least willing to co-operate with the latest development fad, however ecologically sound it happens to be. This situation has major implications for the design and implementation of many protected area schemes, in particular for the types of regulations and incentives, levels of compensation, and forms of participation and dialogue.

The conservationist approach has provoked social conflicts which often undermine the original policy objectives. In such circumstances, non-enforcement of regulations may become an explicit strategy to avoid or reduce tensions. This, coupled with the objective limitations on human and financial resources necessary for the administration of protected areas, has meant that reserve status often exists only on paper.

Chapter 9

The project approach: Reforestation, sustainable logging, agroforestry and social forestry schemes

Growing national and international concern with the problems of deforestation and environmental degradation have been translated into hundreds of projects and programmes involving reforestation, the promotion of sustainable forest management practices, agroforestry and social forestry schemes. Numerous NGOs, bilateral and multilateral aid agencies, corporate enterprises, rural producers and landowners, as well as state agencies, are involved in this field. A variety of concerns and interests underlie the involvement of such agents: conservation; developmentalist or macro-economic concerns which see the lumber industry as playing a positive role in national development and the balance of payments; private interests which see the potential for profitable investment in trees and forests; and the basic needs concerns of peasant farmers and rural community residents. This chapter examines the experience of implementing projects and programmes of this nature.

REFORESTATION

Until the 1990s, relatively little had been done in most countries to encourage reforestation and sustainable forest management practices. In 1980, forest plantations in the region accounted for just 25,000 ha. By 1987 there had been only a slight increase in this figure (Hedström, 1990:58). While annual deforestation accounted for approximately 340,000 ha in the late 1980s, the annual rate of reforestation was approximately 20,000 ha, much of this in one country.

In Honduras, where deforestation accounted for 80,000 ha annually during the mid-1980s, government reforestation programmes replaced just 2,000 ha annually (Hernández, 1986). While this ratio of 40 to 1 was probably the highest in the region, in other countries, the area deforested each year vastly exceeds that which is reforested. The ratio in Nicaragua was reported to be 20 to 1 in 1990 (*Revista del Campo*, 12 January 1990). Despite the scale of the ecological crisis in El Salvador, it was estimated that reforestation accounted for just 1,000 ha a year and that the total area which had been reforested by 1990 amounted to no more than 14,000 ha (Mansur, 1990:22).

Even in Costa Rica, a country often portrayed as a forestry protection show-piece, it was reported in 1982 that the ratio of deforested to reforested land was 12 to 1 (see Brockett, 1990:91). Between 1980 and 1990, total national reforestation for commercial production amounted to approximately 40,000 ha (Bradley *et al.*, 1990a:70). The annual rate of reforestation was expected to increase to slightly more than 20,000 ha during the early 1990s (Current D., 1991:3; *Tico Times*, 7 June 1991:5).

Since the late 1980s, the rate of reforestation in Costa Rica has increased as a result of various credit and fiscal incentives. The 1986 Forestry Law provided greater incentives for reforestation on private lands – both large landholdings and peasant farms – by introducing transferable bonds for reforestation projects which can be used to pay taxes or sold at slightly less than face value (Bradley *et al.*, 1990a:70–71). It was expected that during 1988 and 1989, approximately 20,000 ha would be reforested under the fiscal incentive scheme (MIRENEM, 1990). By 1990, however, virtually no policies existed for reforestation on public lands (Bradley *et al.*, 1990a:72).

As indicated in the previous chapter, stricter controls were also introduced on the use of forest resources on private lands. The Costa Rican government decreed a state of Forestry Emergency in September 1987, in order to halt deforestation. The decree sought to ban credit for agricultural activities which involved the removal of forest, oblige the agrarian reform institute IDA to set aside 10 per cent of IDA settlement farm plots for forestry development, and ban the export of unfinished forest products (ibid.:65). Particularly important for curbing rates of deforestation has been the fact that fewer incentives are now being provided for cattle expansion while ranchers are encouraged to convert pasture to forest.

The Guatemalan government also introduced a fiscal incentive programme which enabled companies or individuals to obtain tax concessions of up to 50 per cent on their taxable income (DIGEBOS, 1990). While such schemes seem attractive on paper, they have experienced considerable limitations. The area reforested under the Guatemalan programme has increased somewhat in recent years but still only accounted for approximately 1,000 ha a year in 1990. The private companies involved in promoting and executing the programme found that their market was somewhat limited given the tradition of tax evasion which exists and the unwillingness of many firms to risk channelling funds into long-term investments during times of economic difficulty. Under these conditions it appears that the companies and individuals prepared to take advantage of fiscal incentives are often the large conglomerates and their executive directors. Several United States transnational corporations have been particularly interested in the programme. Participation in such schemes can, of course, be extremely useful for publicity purposes. It is common to see in Guatemala, media advertisements which show how such and such a company is saving the country from environmental destruction. What is not mentioned, however, is that the law permits those participating in the scheme to completely clear all the reforested area after just nine years.[1]

Legal aspects governing natural resource property rights can also act as a disincentive. In certain countries such as Honduras and Nicaragua it is the case, or has been until recently, that forests on private property belong to the state. Reforestation schemes which fail to provide guarantees regarding ownership rights are likely to receive a lukewarm response. Moreover, landowners and companies participating in such schemes generally confront a growing body of regulations governing tree felling. In Nicaragua, for example, there was, until 1992, little incentive for private producers to undertake reforestation. The Forestry Law of 1980 nationalized the country's forests. In 1990, the situation was such that private landowners wanting to fell more than 10 cubic metres had to obtain a permit from the Institute of Natural Resources (IRENA) and agree to pay the government the cost of the wood. They could retain approximately one-third of the value. Hence for every cubic metre of standing pine valued at approximately $15 in 1990, the producer received around $5.[2]

Major problems have arisen regarding the quality and sustainability of reforestation schemes throughout the region. Referring to tree plantations in Costa Rica, Sargent notes that 'success should not be measured only in terms of the area planted. Serious problems have affected the quality of reforestation activities . . .'. She cites, for example, the use of inappropriate tree species or mismatch between site and species; the frequent use of poor quality seed; the lack of demand for plantation products; and the high costs of logging and transport of products to processing centres, when reforestation occurs on relatively small blocks dispersed over a large number of holdings (Sargent and Bass, 1992a:152–3).

Also relevant to the question of quality is another problem which has affected plantation schemes in the region, namely the clearance of natural forest or woodland to make way for monoculture tree planting. Hence, there has often been a trade-off between increased timber production on the one hand, and biodiversity or soil regeneration (associated with fallow) on the other hand.

In several countries funding has been available for planting but not for management practices involving caring for saplings or young trees. This was the case in El Salvador during the early 1980s when funding from USAID led to areas of between 50 ha and 100 ha being planted with trees on many of the farms that formed part of the agrarian reform programme. In all, a total of 8,400 ha were planted. The initial funding, however, included nothing for management and many of the trees planted during 1980 and 1981 perished.[3] A top-down approach to project design and implementation also led to serious problems. Many of the intended beneficiaries failed to see what benefits could be gained from planting trees on land that could be used for food or other crops. Animosity towards the project was such that instances arose where local farm residents set fire to areas recently planted with trees.[4]

Forestry technicians consulted in Totonicapan, Guatemala, estimated that, until the late 1980s, up to 95 per cent of the trees planted in reforestation projects perished due to problems of management and extension services. These included the lack of transport (for extensionists, deliveries of seedlings, etc.), limited knowledge regarding management practices, the high incidence of forest fires, uncontrolled grazing in reforested areas, and the low motivation of local people participating in such schemes. Seedlings, for example, would be planted without removing the bag, neither were they protected from frost. Those that survived were often smothered by weeds. As a result they developed poorly and ran a greater risk of being destroyed by fire. Reforestation projects in the zone also faced the problem that producers would tend to abandon silvicultural practices at the beginning of the sowing season which happened to coincide with the period when more care of seedlings was required due to extreme climatic conditions.

SUSTAINABLE FOREST MANAGEMENT

A particularly problematic item in the portfolio of activities associated with forest protection concerns efforts to promote so-called sustainable logging or 'sustained yield management'; that is, management for sustained timber production (Poore, 1989:191). Interpretations of what this actually involves vary considerably. For some it means that the volume of timber extracted in a period of years should not exceed the volume of new growth. Others, such as Poore, argue that:

> ... the single most important condition to be met is that nothing should be done that will irreversibly reduce the potential of the forest to produce marketable timber – that is, there should be no irreversible loss of soil, soil fertility or genetic potential in the marketable species. (ibid.:5)

It is apparent that very little has been achieved in this area. In fact the Central American situation resembles that found throughout much of the world's tropical forest regions. The conclusions of an IIED study on sustainable forestry in member countries of the International Tropical Timber Organization (ITTO) seem highly relevant for Central America. While there is increasing awareness among foresters and policy makers in many countries that action is needed, and although some steps are being taken to promote sustainable logging:

> ... [n]evertheless, progress in establishing stable sustainable systems is still so slow that it is having very little impact on the general decline in quantity and quality of the forest....
>
> The extent of tropical moist forest which is being deliberately managed at an operational scale for the sustainable production of timber is ... negligible. (Poore, 1989:xiv–xv)

Referring to the Latin American and Caribbean ITTO countries where case studies were conducted for the IIED study (which included Honduras), Synnott writes:

> ... the following components are generally weak or lacking: advance planning of the location and intensity of the annual cut; supervision and control to ensure that the cutting conforms to the planning; and protection of the area to limit unplanned activities including settlement and uncontrolled logging. (Synnott, 1989:75–6)

A regional workshop on the sustainable management of lowland broadleaved forests, organized in 1990 as part of the Central American Tropical Forestry

Action Plan, analysed some of the experiences in this field (Plan de Acción Forestal para Centroamérica, 1991). Apart from being few and far between, all the experiences referred to had only recently commenced, dating from the late 1980s, and had yielded few, if any, concrete results.

Although more has probably been achieved in Costa Rica than in any other Central American country, a 1990 USAID report was forced to conclude that effective incentives for forest management had yet to be defined:

> Aside from certain tax breaks, no other incentives in the Forestry Law stimulate and improve forest management. The absence of incentives for forest management similar to those for reforestation inhibits efforts to improve silvicultural practices in Costa Rica. (Bradley *et al.*, 1990a:87)

For reasons which have been pointed out elsewhere both the concept and methods of sustainable logging must be handled with great care. Some environmentalists maintain that there can be no such thing as sustainable logging since the removal of trees from rainforest areas, at whatever pace, will necessarily impair the ecological functions of forest, leading, for example, to loss of soil quality and habitat (Colchester, 1990b). While this may appear a somewhat purist position, it is certainly the case that the concept of sustainable yield forestry has been transferred somewhat uncritically from north to south – from the relatively simple, less fragile contexts of temperate forests in Europe and North America, to the more complex ecosystems which characterize forests areas in the humid tropics. This northern bias is also evident in the fact that foresters have generally focused on timber as the primary produce of forests and underemphasized the role of other non-wood forest produce which can play an important role in sustaining local employment, in reducing pressures for forest clearance and in encouraging local people to preserve forest areas.[5]

Until fairly recently, the discussions and planning processes concerned with promoting sustainable logging tended to emphasize technical, administrative and financial aspects. These include, for example, financial incentives to logging companies, longer-term concession periods, lack of experience in management techniques, and limited resources to enforce regulations. Increasingly, however, the crucial question of how to integrate local communities in forest management schemes is being posed. But actually promoting local participation can prove extremely difficult. One project which has adopted this approach is the BOSCOSA project that operates in the Osa Peninsula in Costa Rica. The project has sought to establish 'blocks' of forest, ranging from 500 to 5,000 ha, under the administration of local groups. This has involved encouraging families with 'family forests' to associate with others to form 'communal forests'. Progress has been extremely slow, however, due in large part to disputes and complications regarding land tenure (Plan de Acción Forestal para Centroamérica, 1991).

Even imposing basic regulations on logging companies to reduce levels of timber extraction, regulate which trees are removed, and ensure adequate rates of reforestation is likely to prove extremely difficult. This is partly because legal and administrative criteria regarding what constitutes sustainable management remain lax, and partly because enforcement is so difficult.

In the Central American context, however, and no doubt in many other Third World countries, there is a very real danger that the concept of sustainable logging, as applied in practice, could become an excuse to expand logging operations and increase net deforestation. What we have seen in recent years are companies requesting large logging concessions on the assumption that they will manage the forests 'sustainably'. In the absence of effective guidelines to ensure that appropriate plans are drawn up, and mechanisms to ensure that they are implemented, there is a serious risk that a second tier of forest exploitations, purporting to employ sustainable management practices, will be added to the existing tier of more traditional 'mining' operations. As noted in Chapter 2, transnational companies waving the banner of sustainability actively sought lumber concessions in Nicaragua following the change of government in 1990. In Honduras, a large United States company, the Stone Container Corporation, put forward a plan to exploit 216,000 ha in La Mosquitia, one of Central America's largest remaining areas of dense forest (Pickles, 1992:12). While pressure from environmental and other groups or individuals obliged the governments of these countries to turn down these particular requests, it remains to be seen whether they will be able to resist the ongoing efforts of the transnationals to expand their logging operations in the region.

AGROFORESTRY AND SOCIAL FORESTRY SCHEMES[6]

The vast majority of the region's agricultural producers are small peasant producers located in hilly areas prone to erosion and often farming land more suited to forest or perennial crops than annual food crops. This combination of conditions would seem to provide considerable potential for the promotion of agroforestry systems among small farmers. While several agencies have been actively working in this field since the 1970s, the adoption of new agroforestry systems by small farmers has proceeded relatively slowly. This section examines the types of constraints which account for this situation.

The promotion and adoption of agroforestry systems involves a complex chain of activities involving experimentation, research, training, extension, production and marketing. Some notable advances have been made in several of these components. This has been due in no small measure to the work of CATIE which, since the 1970s, has assumed a leading role in agroforestry research in Latin America.

CATIE's initial research programme sought to identify traditional agroforestry systems employed by producers in the region. It found that many of these systems were circumscribed by local ecological conditions and not easily transferable from one setting to another. After 1980, attention turned to the promotion of exotic species and research into new agroforestry systems, involving tree–crop or tree–pasture combinations considered appropriate to the different agro-ecological and socio-economic conditions of the region.

For several years, however, the focus of this research was very much on agro-ecological and biophysical aspects. Given the scale of fuelwood requirements in Central America and the perceived impact of fuelwood gathering on the state of the region's forests, CATIE's work in the agroforestry field during the late 1970s and early 1980s, was concerned particularly with the identification of quick-growing species of trees for fuelwood. Research focused also on the introduction of nitrogen-fixing trees in crop and pasture areas, live fences, tree species suitable for alley-cropping of maize and beans, as well as suitable tree densities in coffee and cocoa plantations. By the late 1980s, the performance of 150 tree species had been studied. Of these, 24 were identified as having considerable potential in agroforestry systems. CATIE's dissemination strategy currently prioritizes 14 species.[7]

A major aspect of CATIE's dissemination strategy has been the establishment of demonstration plots and farms in collaboration with local producers. By 1990, CATIE had established approximately 30 such farms throughout the region. An estimated 3,000 research plots with between 500 and 1,000 trees have also been set up.[8] CATIE is also actively engaged in training and teaching.[9] This educational process has had a multiplier effect throughout the region. Many people who have received training in CATIE have subsequently set up courses, experimental plots, and projects. Several past students, researchers and teachers at CATIE have also attained influential positions within government and have been instrumental in promoting public policies which favour environmental protection.

CATIE's work in the agroforestry field has been supported mainly by USAID. A number of strict conditions imposed by this agency have restricted the geographical scope of CATIE's activities. Nicaragua, for example, was excluded from the agroforestry programme during much of the 1980s. USAID refused to fund projects in that country given the United States administration's opposition to the Sandinista government. Restrictions were also imposed on CATIE's activities in Panama when Noriega was in power.

Apart from the work of CATIE numerous other projects are currently underway throughout the region supported by international aid agencies and national NGOs. International agencies such as CARE, GTZ, FAO, USAID, the Peace Corps, CSD and SAREC are particularly active in this field. The central objective of most projects has been to assist small and medium-sized producers. Initially, project participants and beneficiaries consisted mainly of individual producers who were willing to collaborate in experimental projects. In recent years, however, there has been an effort to involve a greater number of producers, with attention being focused not only on private producers but also co-operatives, communities and local committees. Working with community residents and small groups of farmers has facilitated the tasks of technological diffusion and environmental awareness-raising.

More than two decades of research into agroforestry systems has produced a voluminous body of findings and literature on the biological and, to a lesser degree, the economic and socio-economic advantages and disadvantages of such systems (Budowski, 1981).[10] It was not until the latter half of the 1980s that the importance of socio-economic and socio-cultural aspects was recognized more fully and that these elements were incorporated into

CATIE's research and training programme. Such aspects, however, have clearly remained one of the weak links in the agroforestry programme, particularly at the level of outreach activities.

This has been partly due to a certain disciplinary bias which has characterized research in this field. It also reflects the fact that it is only recently that new agroforestry systems have been promoted among a significant number of producers. Most studies have tended to examine the motivations of producers for accepting or rejecting agroforestry systems or particular tree species and the priorities of producers *vis-à-vis* specific forest products.

It is still difficult to gauge precisely the economic benefits associated with agroforestry systems. Economic analysis of agroforestry systems is at a preliminary stage, partly due to the fact that reliable (ex-post) information on production costs, market prices and yields has only recently become available. CATIE is currently processing information from demonstration farms which should facilitate economic analysis. Some preliminary ex-ante cost-benefit estimates indicate fairly positive rates of return which generally exceed alternative rates (that is, bank interest rates) by a significant margin.

Table 9.1 Preliminary financial assessment of selected agroforestry associations in Central America

Species	System	Establishment Cost ($)	Person-days	Type of product	Cash flow period (years)	Projected rate of return (%)	Discount rate of return (%)[1]
Bombacopis quintatum	Plantation	285	32	Sawn wood, construction posts	26	12	12
Cupressus lusitanica	Plantation	373	46	Sawn wood	24	10	8
Eucalyptus with Zea mays	Taungya	357	124	Fuelwood	4	20	10
Gliricidia sepium	Plantation	493	68	Posts and fuelwood	4	28	8
Gliricidia sepium	Live fence	260	21	Live posts, forage	20	34	12
Coffee, Cordia & Erythrina		ND	ND	Coffee, timber	20	34	12
Guazuma ulmifolia	Plantation	392	63	Fuelwood, posts, forage, sawn wood	24	25	8
Acacia mearnsil	Protection	405	50	Fuelwood, protection of soil	18	28	10

Source: Based on Reiche, 1988, Table 5

1. Interest rates in different countries.

It is apparent, however, that returns vary considerably by agroforestry system. The data presented in Table 9.1 refer only to direct costs and benefits. Many other indirect costs and benefits are associated with agroforestry systems which should be incorporated in such an analysis. Certain analysts believe that the indirect benefits (reduced erosion, nitrogen-fixing, physical security to property, shade and so forth)[11] probably outweigh the indirect costs. But (at the time of writing) this type of analysis had yet to be attempted.

Other question marks exist, however, regarding cost-benefit analysis. It would seem that little attention is being paid to the conceptual problems involved, not least the problem of what value to put on so-called 'intangibles' or goods and services for which there are no markets (shade, cultural or aesthetic perceptions of trees, etc.). Moreover, by attempting to reduce values to the same common denominator, cost-benefit analysis has difficulty in dealing with the fact that different social groups (for example, subsistence peasants and large commercial farmers) have very different needs, beliefs, priorities, preferences and, therefore, ways of assessing 'cost' and 'benefit'.[12]

Other more practical problems affect cost-benefit analysis. Much of the analysis of opportunity costs appears to remain at the level of what a specific amount of money would have earned had it been placed in a bank deposit account (or commercial interest rates on agricultural loans), rather than in terms of the additional income a farmer might have earned by undertaking an alternative on-farm investment. Moreover, while the process of selecting a demonstration farm (from which data is derived) usually takes into account the question of whether or not the producer involved is representative of farmers in the area, it is often the case that the ease with which the selected farmers subsequently obtain certain inputs and services (notably technical assistance) renders them atypical.

Both reforestation and agroforestry schemes have had important implications for the gender division of labour and other social relations. They have provided employment opportunities for women and children. In Honduras, agencies involved in setting up nurseries for reforestation and agroforestry schemes often prefer to employ women workers whose handling of seedlings tends to be better. In the community tree nurseries programme in El Salvador (discussed in Chapter 10) women and children make up 50 per cent of the workforce. Women's participation in such schemes has also enabled them to acquire certain silvicultural skills which can be applied on the farm.

In some areas, though, changes in the gender division of labour arising from women's participation in reforestation and agroforestry schemes have been opposed by men. A study of the role of women in reforestation projects in Hojancha, Costa Rica, found that very few men considered that women could or should plant trees (Major *et al.*, 1989). Of the men interviewed, 41 per cent believed that the women's household duties restricted involvement in such work. However, only 7 per cent of the women interviewed believed this to be the case. Women generally cited poor health (29 per cent), old age (29 per cent) and lack of time (21 per cent) as being the principal constraints on participation. It was found that where women participated in tree planting activities they not only worked well but were able to accommodate such work with household activities.

Many development planners and practitioners agree that women's participation in such projects has been highly beneficial. This phenomenon, however, clearly needs to be studied with a more critical eye than has been the case to date. At least three questions immediately come to mind. First, to what extent is the drive to involve women and children in such schemes related at all to differential wage rates, that is, the fact that they are often prepared to work for lower wages? Second, how does involvement in such schemes affect the workload of women or the situation of children? Thirdly, how does women's participation affect the gender division of labour and social relations of production within the tree nursery or out in the field? Put another way, does the participation of women in such activities serve to reduce the workload of men? My suspicions on this latter score were aroused when visiting a large tree nursery in Nicaragua. On arriving at 3 pm I found the three male workers having a siesta while a dozen women and children were out working, hunched over their seedlings, under a blazing sun.

With some notable exceptions (some of which are discussed in the following chapter), tree planting projects and programmes have tended to remain as somewhat isolated experiences, or have been concentrated in specific regions of a particular country. Making the qualitative leap from model farms and demonstration plots to widespread diffusion of agroforestry systems and techniques among peasant and commercial farmers has proved extremely difficult due to numerous biophysical, social, economic, cultural and institutional problems. Despite the considerable efforts which have gone into research, experimentation, training, and establishing extension systems, the number of producers who have actually adopted the new technology is still very small.

The search has been on in recent years to find the key elements which account for the success or failure of agroforestry and social forestry projects. A number of evaluations have identified a range of aspects associated with institutional constraints and biases, questions of incentives and participation, as well as factors crucial to the continuity or sustainability of projects.

Institutional constraints and biases

Various studies reveal that the central problem restricting the diffusion of agroforestry systems is not one of farmers' ignorance regarding the ecological and economic functions of trees. Many producers are conscious of the phenomenon of environmental degradation and its effects on production and productivity and recognize the need to act on this front (Dittborn, 1988; Heckadon, 1990; Reiche, 1988). The reluctance of producers to introduce or expand agroforestry schemes generally derives from a combination of social, legal, institutional, administrative and economic constraints.

While the adoption of new agroforestry systems has yet to 'take off', the experience to date has yielded important lessons as to what does and does not work. Particularly important are questions relating to the size of farms, tenure, and access to finance. The producer may also lack vital information regarding how to obtain or use the new technology, as well as the marketing of forest products.

Many of these factors have to do with structural conditions which characterize peasant production in the region and the nature of the land tenure system. Producers on relatively small plots – concerned primarily with investing their time, energy and limited finance in the production of crops which provide the food and income necessary for subsistence provisioning, and who often lack secure claims on the land they farm – are often reluctant to plant trees (Dulin, 1985; Dittborn, 1988; Jones, 1982).

As indicated in Chapter 3, insecurity of tenure is a particular problem in agrarian frontier or forest areas where colonization has occurred. It often serves not only to encourage deforestation but as a disincentive to the adoption of agroforestry systems or forest management practices. As pointed out in one study of employment generation in agroforestry zones in Costa Rica:

> In the three [communities] studied, the producers do not possess property titles. While this has not posed an obstacle to the exploitation of forest resources, it generates ... considerable uncertainty regarding length of residency and the possibilities of managing forest resources as a source of income for family subsistence. (Mora, 1991:48 – author's translation from Spanish)

Insecurity of tenure may pose an obstacle to tree planting and management but it cannot be assumed that simply granting property titles will prompt significant changes in natural resource management practices. Much depends on what happens to the producer's or rural family's access to resources such as finance, inputs, and technical assistance, as well as market outlets for tree products. It will also depend on the incentives available for other primary products and the availability of labour for tree planting and management activities.

The limited number of agricultural producers involved in agroforestry schemes is also related to serious limitations which affect the work of agencies providing finance and support services. Technical assistance and extension services have a crucial role to play in the promotion of agroforestry systems. As Tschinkel observes, one of the key elements for the successful promotion and continuity of agroforestry systems consists of 'very personal, intensive extension service with frequent follow-up visits until the farmer has achieved confidence in his tree crop' (Tschinkel, 1984).

The personal and intensive character of technical assistance should also, however, have to do with the need to tailor the diffusion of agroforestry practices to local needs and conditions. Many agencies have committed the error of imposing 'models', or generalized prescriptions, in rural settings where considerable heterogeneity exists regarding livelihood systems, social structure, and agro-ecological site conditions (Mora, 1991). Moreover, in many rural areas, producers already practise certain, so-called 'traditional' forms of agroforestry. Very often project success will depend upon the capacity of agency personnel and extensionists to build on this existing body of knowledge. As Mora observes:

> In such cases, it is not a question of starting from scratch or of transferring a model which is not known to the producers. Rather, it should be a case of strengthening systems which have been developed by the producers, whose efficiency has been demonstrated or which can be easily evaluated. (Mora, 1991:53 – author's translation from Spanish)

Relevant in this context is the controversy surrounding the promotion of exotic tree species such as eucalyptus, as opposed to indigenous species. Many agencies, including CATIE, have focused their attention on the former.[13]

Serious problems have arisen with state agencies in various countries. These include the unwillingness of agencies to commit significant resources to agroforestry due both to budgetary constraints and the lack of a clearly defined forestry and agroforestry policy. Extensionists are often unmotivated or unable to work in the field given restrictions on transport, gasoline, and funds to cover field expenses. It is a somewhat pathetic sight to tour a region and find numerous projects or extension services which have virtually ground to a halt because of shortages of counterpart funds and supplies. The words of one agroforestry project adviser in Costa Rica could apply virtually anywhere in the region:

> Often nothing gets done because if a motorbike is available, there is no fuel. If fuel is available there is no bike, and when both are available, there is no petty cash for field expenses.[14]

Many of the constraints associated with technology transfer identified in a study of agroforestry projects in Honduras are relevant for much of the region (Dulin, 1985). Producers or groups involved in local schemes often have to obtain seedlings from central nurseries located considerable distances from the farm. Also, it is common for agencies involved in agroforestry projects and programmes to attempt to obtain the collaboration of producers by relying on verbal explanations instead of the far more effective alternative of visits to demonstration plots.

CATIE does not actually provide extension services but liaises with extension agencies in the different Central American countries and provides them with training and documentation. CATIE's capacity in this regard, however, is fairly limited. In Guatemala, for example, only 17 extensionists participate in the programme, although some 76 peasant 'promoters' are also involved. In Honduras, where CATIE works with a relatively large number of extension agencies, only 60 extensionists participate in the programme. The principal extension agencies promoting the use of agroforestry systems assisted just 214 producers in 1990. CATIE cannot satisfy the demand for training which exists in the region. The Honduran Coffee Institute, for example, recently requested training for most of its 130 extensionists. CATIE currently works with only six of the Institute's extensionists and is under pressure to expand its programme.[15] The problems which affect the operations of state agencies have convinced CATIE of the advantages of working more with non-governmental organizations and to encourage organized communities to assume more responsibility for the promotion of agroforestry schemes.

Extension services have become particularly vulnerable in the current age of budget cuts associated with economic stabilization programmes. Extensionists are often among the first to lose their jobs or find their operational funds slashed. Technology diffusion programmes have also been affected by the rapid turnover of personnel in state agencies as extensionists and foresters abandon their jobs due to low pay. There is often a rapid turnover of skilled agency personnel. This may be due to the fact that many trained personnel leave as a result of problems of low pay or delays in pay when funds are channelled through the central state apparatus. Instability in employment also arises given the nature of the project cycle. While effective agroforestry programmes require a long life, specific projects tend to cover a relatively short time period of just a few years. Second or third phases may be approved to give continuity to an original project but such approval often comes just as an earlier phase is finishing or even after it has terminated. In the meantime, many project personnel may have already sought employment elsewhere.

Women can be particularly affected by limitations on extension work or the gender bias which characterizes the work of many extensionists. A study of women's participation in reforestation projects in Hojancha, Costa Rica, revealed that lack of information and technical assistance were two of the principal elements which restricted their involvement in agroforestry and reforestation schemes (Major *et al.*, 1989). While women showed considerable interest in such schemes, their experience had largely been restricted to tending fruit trees in the household plot. They were not invited to meetings or 'field days' to work with extensionists. In an area visited frequently by extensionists, only 10 per cent of the women interviewed had received information about agroforestry from extensionists. When women were asked what would enable them to participate in such schemes, 60 per cent referred to the need for more information and attendance at meetings.

The resource constraints affecting many state agencies involved in reforestation and agroforestry activities reflect to some extent the bias which generally exists in favour of commercial agriculture. While in several countries there has been more support during the past decade for small farmers, the agencies involved tend to prioritize purely agricultural, usually export-oriented activities, such as the production of so-called 'non-traditional crops' (vegetables, fruit, flowers, spices, etc.). Moreover, banks are often unwilling to provide credit for tree planting and maintenance, or agroforestry products. As a result such activities generally have to rely on project funding from foreign aid agencies.

The government of El Salvador has recently attempted to correct this bias by urging the country's agricultural extension service to promote agroforestry systems. The problem remains, though, that such services reach only an extremely small percentage of farmers. In 1988, it was estimated that just 6.5 per cent of the country's farmers received technical assistance (Nuñez *et al.*, 1990:5).

Even when agroforestry projects have been approved and are being implemented, the agricultural bias of ministry officials can pose an obstacle to project execution. Extensionists working on the community nurseries programme in El Salvador, for example, found that bureaucratic delays in obtaining supplies of essential inputs had arisen due to the fact that agronomists who still regarded forestry activities as secondary, controlled the administrative process through which supplies and funds were allocated (Heckadon, 1990:61).

The limited scope of agroforestry schemes may also be related to the tendency of certain agencies to work with larger farmers. This may be due to political reasons or, alternatively, reflect the fact that agencies often set goals and evaluate their performance in terms of variables associated with the area and number of trees planted, rather than numbers of farmers who participate in such schemes. One study of seven projects among small hillside farmers found this to be the case in Honduras (Dulin, 1985). As pointed out by Current, referring to an agroforestry project in El Salvador:

> This emphasis on area often has the effect of pushing field people towards larger landowners to be able to meet their goals and thus tends to bias extension efforts towards larger landowners at the expense of smaller landowners. (Current D., 1991:49–50)

In a country such as Costa Rica, the implicit bias which characterizes extension services is not so much towards large farmers as small or middle-sized commercial farmers. This was found to be the case, for example, in one study of farming systems in an area of Puriscal county where a relatively successful agroforestry project, involving coffee and shade trees, had been implemented (Brüggemann and Salas, 1992). The extension services had largely bypassed poorer peasant producers. But this situation should not be explained purely in terms of the preferences of project personnel and extensionists. As occurred in this particular area, it is often the case that those receiving the bulk of the goods and services associated with specific projects and programmes, happen to be those who are organized and able to influence patterns of resource distribution. In certain countries, notably Costa Rica and Nicaragua, it is apparent that the power and influence of this stratum of rural producers increased considerably during the 1980s.

Incentives and participation

Many agroforestry and social forestry projects have relied on fairly inappropriate material and financial incentives. This has been particularly apparent in projects using food-for-work incentives. As Dulin observes: 'Suddenly, the producer begins to feel like a wage worker employed by the programme and becomes reluctant to take the initiative unless provided with an incentive' (Dulin, 1985).

The question of incentives is intimately tied up with that of 'participation'. How to secure the enthusiastic involvement of producers and community residents in reforestation and agroforestry schemes, and how to sustain that involvement through time, are crucial questions in this field.

Many projects have attempted to impose blueprints or formulas regarding tree and crop combinations which prove ineffective when applied in practice. The assumption, for example, that everyone in rural areas uses and needs fuelwood has led certain agencies to promote the use of trees for fuelwood even in areas where access to fuelwood is not a major problem. Not surprisingly, the farmer's response has often been lukewarm. The failure to consider sufficiently past and future land-use practices and the priorities of the producer regarding cropping systems is another problem which has impeded the implementation of projects.

Closer contact with farmers has led to a number of changes in project approach. Notably, it has led to a recognition of the need to promote more diversified agroforestry systems involving tree species capable of producing a variety of products, including not only fuelwood but also stakes, posts, poles, food and timber for construction.

There are some indications that, through time, a more participative approach to experimentation and extension has emerged. Extension agencies with which CATIE liaised in the different countries, for example, were encouraged not to dictate to farmers which species they should plant. The so-called 'demonstration effect' came to be regarded as one of the most effective means of encouraging farmers to incorporate trees which CATIE considered most appropriate into their farming systems. Local farmers must be able to see for themselves concrete results in a relatively short period of time. For this reason certain species of eucalyptus have proved particularly popular in many agroforestry projects despite the often publicized disadvantages associated with them in certain ecological and socio-economic contexts.

A relatively new experience which has yielded some promising results in countries such as Honduras and Nicaragua has been the use of peasants to teach peasants. Referring to a soil conservation project in north-western Nicaragua, Cardenal writes:

> During [the late 1980s], a *campesino*-to-*campesino* training system was instituted, in which the instructors themselves were *campesinos*. The co-operatives that had more experience and had taken part in training workshops taught their companions different techniques, such as how to . . . dig furrows on a graded curve, how to make terraces and filtration ditches, and how to run a vegetable or tree nursery. They also discussed such issues as the best seasons for planting and the more useful species. In short, the training system imparted a collection of methods and information related to conservation and land management that allows the *campesinos* to resolve their own problems. (Cardenal, 1992:70)

This approach is important for it not only facilitates communication and dissemination but also ensures that new practices and techniques draw on local knowledge and are tuned to local needs and capabilities. Moreover, through such an approach, the concept of 'participation' is not restricted to simply ensuring the enthusiastic involvement of individuals and groups in a specific project or programme. Participation is also about the capacity of local groups, particularly disadvantaged ones, to influence decision-making and planning processes that affect their lives. While development agencies throughout the region have clearly become more aware of the need for participation, their definition of what participation means is generally restricted to the former aspect.

Continuity and sustainability

Certain projects and programmes have also failed due to inadequate follow-up. Producers have often lost interest in schemes and failed to maintain or manage the trees they have planted. Once trees have actually been incorporated in agroforestry systems the problem of 'sustainability' arises in relation to both biological aspects of tree growth and social aspects of improved welfare for local farmers and community residents.

How to ensure that agroforestry systems will be successful and continue to expand following the termination of a specific project is a complex question. A recent study which evaluated the performance of 12 forestry projects, programmes and land-use policies throughout the region identified some of the major aspects which have a bearing on sustainability and which should be considered carefully by planners and policymakers working in this field (ibid.).

It is crucial that basic technical knowledge is passed down both to extension agents and local residents and that the technical packages used are relatively simple. Here the question, raised above, of appropriate extension methodologies is particularly important.

Counterpart agencies which continue to be involved in agroforestry and reforestation schemes once a donor agency has withdrawn, must have the capacity to maintain equipment which has been acquired. Whereas many state and donor agencies are more than willing to acquire new machinery and equipment, they often fail to prioritize maintenance activities. Maintaining the existing stock of equipment becomes even more difficult when the use of 'tied aid' means that national agencies have to import goods made in a specific country. This can result in a small country having to find the spare parts and expertise to maintain various makes of what is essentially the same product.

Continuity also requires secure outlets for tree products. Research related to the promotion of agroforestry systems has tended to focus on so-called 'upstream' activities (production of seedlings, transplanting and tree or forest management). It is crucial, however, to consider the entire chain of activities which characterizes the forestry sub-system, including both the processing and marketing of wood products.

As one might expect, the attention of agroforestry extensionists and agencies has focused on how to get suitable trees incorporated in local farming systems. Many producers, however, who have participated in such schemes are now finding themselves with considerable quantities of wood products to dispose of. In some areas or countries, notably El Salvador, scarcity and high demand for such products has made marketing relatively easy. In many other areas, however, finding marketing outlets is not so straightforward. This aspect, for example, has caused problems in certain areas of Costa Rica and Honduras. The ongoing interest of small farmers in agroforestry systems may also be affected as a result of the vulnerable market position of peasant farmers and the nature of merchant/peasant relations which leaves the latter receiving a minimal price for their products. As Mora observes when referring to certain communities in Costa Rica:

> The lack of financial resources, equipment and means of transport, means that the extraction of trees cannot be carried out by peasant families. In general, the merchants buy stumpage (standing trees) and thereby obtain the bulk of the benefits derived from forest products. (Mora, 1991:49 – author's translation)

Following the chain even further, it is also pertinent to consider the question of what farmers or forest dwellers will do with the additional income derived from the use of forest resources. How will the way in which this income is used affect farming or resource management systems, and what are the environmental implications of any such changes? The possible contradictions which may arise can be illustrated by way of an anecdote. A project supported by CATIE to promote the sustainable use of forest resources carried out a feasibility study in the Bocas del Toro province of Panama to determine which forest products could be exploited by the local Indian population on a sustainable basis. The study revealed that prospects for the production and marketing of heart of palm (*palmito*) were good. The question was put to the intended beneficiaries as to how they would use the increased income derived from this production. The reply of 'expand cattle production' (and, by implication, clear forest for pasture) was not what the environmentalists associated with the project expected or hoped to hear.[16]

There is a dearth of information and research regarding the social impact of introducing or expanding agroforestry systems. 'Sustainability' is often viewed in terms of tree development through time, rather than with socio-economic development which increases human welfare through time. It is often assumed that tree development will necessarily bring with it tangible net benefits for the people involved – but this assumption cannot be taken for granted.

While it is apparent that numerous projects and programmes promoting reforestation, sustainable forest management and agroforestry systems have been and are being implemented throughout the region, it is extremely difficult to assess what the environmental and social impacts have been. One of the weakest links in the project chain relates to follow-up and evaluation. It is generally the case that projects end once the trees are planted or a certain number of people have received training courses. Whether the trees actually grow and land-use patterns are changed, or whether the knowledge gained by those participating in projects is applied, remains a big question mark.

Despite the fact that Central America is the site of one of the world's most active agencies in the field of agroforestry, it is nevertheless the case that only a relatively small number of farmers have actually taken up the new technology. The Central American experience would appear to indicate that even when a significant commitment exists on the part of governments and international donor agencies, research and experimentation into appropriate species, and designing adequate systems and methodologies of technology transfer can take at least two decades. One is then confronted with the considerable institutional constraints which limit the capacity of extension agencies to reach large numbers of farmers.

The time-scale involved has implications for external aid for agroforestry projects. There is a danger that 'donor fatigue' can set in. Aid agencies generally like to see results in just a few years and may be reluctant to provide long-term financing. Moreover, the bureaucratic procedures normally followed for renewing funding for projects often mean that budget authorization is not acquired until the last moment or even after an earlier phase of a particular project has been terminated. In the meantime, project personnel may have sought employment elsewhere (Current, D., 1991).

The dependency of environmental rehabilitation projects on external aid is an extremely complex issue. Numerous problems characterize donor–recipient relations. Some of the more technical difficulties associated with project design and implementation, as well as the clash of interests and priorities of national and international agencies, have been identified by Green (Green, 1990). Other problems, perhaps more serious, relate to the fact that many national institutions simply do not have the budgets or personnel to fulfil their counterpart obligations.

This was borne out clearly in an evaluation of Swedish aid to the Nicaraguan forestry sector (Budowski and Vieman, 1989:125). Such a situation can also lead to donor fatigue. The same evaluation detected other important problems which characterized the donor–recipient relation: notably, the tendency to transfer uncritically, to the tropical conditions of Nicaragua, knowledge and techniques appropriate to the temperate climate of northern Europe; the insufficient attention paid to sustainable forestry management on the part of Swedish agencies supporting the local forestry industry; and the tendency to draw up over-optimistic plans which failed to consider the complex array of social and economic forces underpinning deforestation (ibid.:125–6).

Chapter 10

Programme and project implementation: Concrete experiences from El Salvador, Guatemala and Honduras

In order to gain a better understanding of the types of socio-economic, cultural and political constraints which affect the implementation of agroforestry and social forestry projects and programmes, I examine, in this chapter, three specific experiences. These include:

- the community nurseries programme in El Salvador;
- the case of small scale agroforestry and reforestation projects in Totonicapán, Guatemala; and
- the social forestry system in Honduras.

COMMUNITY NURSERIES PROGRAMME IN EL SALVADOR

One of the most extensive outreach programmes to be implemented on a national scale is that of the community nurseries scheme in El Salvador. This programme commenced in 1984 and has involved more than 13,000 individual peasant and co-operative producers during the first five years. By the end of 1989, 180 tree nurseries had been established. Evaluations of this programme, carried out by researchers based at CATIE (Current D., 1991; Heckadon, 1990), have identified the following aspects to be particularly pertinent to the success or failure of the different projects.

Successful experiences regarding the establishment of tree nurseries and the participation of farmers in this social forestry programme generally occurred in communities with a certain level of organization and where the project leader was not only active but also acted in the interests of the community at large. Where these characteristics were absent, the projects often failed. The process of selecting project leaders was important. As Heckadon observes, it is crucial that it is the community and not the extensionist that selects the project leaders. The extensionist is often tempted to choose those producers with whom they have a personal or political affinity. Very often the person chosen by the *técnico* will be one of the richest farmers in the area whose communal interests come a poor second to those associated with personal gain (Heckadon, 1990).

The success or failure of certain projects has also been related to the type of relations which exist between farmers or community residents involved in the project and the extensionist. Authoritarian attitudes on the part of the latter or attempts to impose rigid formulas regarding species and agroforestry practices often alienate the local population. Farmers are likely to have their own priorities regarding appropriate tree species and land management practices. The technical guidelines which had been designed originally for the tree nurseries programme, for example, envisaged that trees would be planted in small boskets. Due to the small size of the farms, however, most producers opted to plant in single rows along fences or river banks (ibid.:52). Also, not only in El Salvador but throughout the region farmers have often preferred, for example, fruit trees which did not form a central part of CATIE's research, or cedar, which was not one of the species prioritized by CATIE. Many agencies which are active in the promotion of agroforestry systems have come to realise that it is crucial to respect the priorities of the farmer regarding appropriate species, even if these contradict those of 'the expert'.

There have been several cases where the recommendations of the extensionist regarding appropriate species have proved an outright failure. One of CATIE's 'favoured' species, leucaena, proved particularly problematic due to the poor soils which often characterize peasant holdings in El Salvador and elsewhere. Moreover, it has proved highly vulnerable to pests and climatic extremes.

The community must want to grow trees. Whether they do or not will depend partly on the situation regarding the supply and demand of different wood products. Here the point mentioned earlier about the limited interest in growing trees for fuelwood, particularly in coffee producing regions where shade trees exist, is relevant. Talks, audio-visual presentations, field days and visits to demonstration plots and farms play a crucial role in this respect, enabling the producer to appreciate how trees can be incorporated into farming systems, that tree growth or the harvesting of tree crops can be achieved in a relatively short timespan with certain species, and the ecological functions associated with trees.

The evaluations of the tree nurseries programme in El Salvador stress the importance of involving women as active participants in the projects. Such participation has not only enabled families to better organize their time, but has also provided employment opportunities for the family as a whole and, by reducing absenteeism, greater continuity to the work process in the tree nurseries. Moreover, it was apparent that women and children were more capable of performing many of the delicate tasks associated with the production of seedlings.

The experience with food-for-work incentives generally indicated that this form of incentive proved useful as a means of engaging people's interest in the programme at an early stage – particularly in motivating the very poor, the elderly and women. It then became more of a hindrance than a help, often reinforcing the idea that participants were working for someone else, rather than for themselves. Moreover, administrative problems associated with storage and transport of food products often affected distribution which, in turn, gave rise to ill-feelings on the part of the local population, thus making the work of the extensionists that much more difficult. Intra-community tensions also arose as a result of the way in which food was distributed. Certain project participants claimed that 'favouritism' entered into the calculations as to who got how much.

The social forestry experience in El Salvador has shown that security of tenure is a key element for success. The programme was relatively successful when it involved families of small landowners or those who had received land through a type of 'land to the tiller' law of 1984 which had given titles to peasants who had previously rented or sharecropped land. The programme was far less successful when it involved landless peasants or members of co-operatives which had been formed during the first phase of the Agrarian Reform when farms exceeding 500 ha had been expropriated. Not surprisingly, perhaps, landless peasants only remained in the programme as long as food-for-work incentives were provided. Many co-operative members also demanded food-for-work incentives since they still tended to regard themselves more as salaried workers than as partners in a communal project. Moreover, some producers felt that the agrarian reform was reversible and that at a future date their lands could be taken away.

Regulations governing tree felling also caused concern among project participants. The existing legislation prevented tree felling without permits on forest lands. Project officers were able to bypass this law by arguing that the trees were planted on agricultural lands. Such concerns, however, were likely to intensify in the future as the programme expanded nationally.

Institutional and administrative constraints have often impeded project implementation. In El Salvador, the so-called 'regionalization' or decentralization of the Ministry of Agriculture which gave more power to the regions, had the effect of adding several bureaucratic layers which greatly complicated the task of obtaining inputs. Delays in the arrival of materials caused many plants to die and dampened the enthusiasm of project participants.

Such constraints increased as the programme expanded. While the government encouraged an increasing number of nurseries, and went so far as to propose a fourfold increase in 1989, it did not provide the necessary material resources and technical personnel to back up its plan. As the number of nurseries expanded, resources became increasingly stretched. As a result, certain tasks suffered or were abandoned altogether. Far less time was allocated, for example, to discussing, with the local population, the ecological importance of silvicultural practices or in follow-up activities associated with caring for seedlings.

REFORESTATION AND AGROFORESTRY PROJECTS IN TOTONICAPAN, GUATEMALA[1]

During 1991, four projects, associated directly with forest protection and tree planting, were being undertaken in the Guatemalan department of Totonicapan where one of the UNRISD case studies was conducted. Three of these projects were supported by foreign aid agencies (CARE, FAO, USAID) but all were co-ordinated and administered by DIGEBOS.

The project 'Tree Nurseries and Reforestation' was financed exclusively with central government funds. As a result, funds were channelled through the Ministry of Finance. This caused considerable delays, which in turn greatly undermined the success of the project. Not only were the forestry technicians unable to fulfil their work plans but local communities participating in the project soon lost interest when project personnel failed to deliver goods and services on time.

Delays associated with some activities, however, have been reduced. Since 1989, for example, DIGEBOS has developed a departmental network of tree nurseries and a system for collecting seeds which reduces the dependency of local nurseries on the national seedbank, BANSEFOR, which supplies most of the country's tree nurseries.

Reforestation was carried out by local communities. In general, however, very little care or management was provided for seedlings and trees. DIGEBOS had very few personnel to provide follow-up services and relied heavily on community-based 'agricultural representatives'.[2] Reforestation was more successful in areas near the departmental capital. This was partly due to the ease with which forestry technicians could visit the areas but also to the fact that local communities recognized the importance of reforestation as a means of maintaining the supply of fuelwood to the urban centre.

The problems of tree care also relate to the question of motivation and incentives. DIGEBOS provided seedlings and technical assistance free of charge. No payment was made to the communities but wire for fencing and tubes for improved fuelwood stoves were provided. However, bureaucratic delays affecting the disbursement of funds, sometimes led to long delays in obtaining these materials. As noted above, in such a situation community residents often lost interest and became suspicious of yet another government institution which failed to provide what was promised.

The project 'Protection of Water Sources', supported by funding from USAID, aimed to reforest areas surrounding sources of water which were important for drinking and irrigation. It sought to work with communities which showed a keen interest in reforestation and the protection of water sources. This approach proved fairly successful and in 1990, reforestation of areas of up to 2 ha took place around 30 water sources. The success of this project, however, was measured solely on the basis of the area which was reforested. There was little follow-up to determine the state of newly-reforested areas.

The project was modified somewhat in 1991 when it was realised that the environmental benefits of carrying out reforestation in such small areas were limited. As a result DIGEBOS decided that reforestation should occur in larger areas but involve fewer water sources. In 1991, the project aimed to reforest areas of approximately 5 ha around four water sources.

This new approach, however, proved problematical. It was often the case that land around the water sources was privately owned and used for agriculture. This led to conflicts between the landowners and the project personnel. As a result, attention turned to communal lands where the community was interested in such a scheme. But here too, problems arose. Often those who lived around the water source were not those who were affected by declining water supplies and deforestation. As a result, their willingness to participate in such a project was limited. Problems of participation in the project arose particularly in communities where much of the population had lost touch with agriculture and the land, and engaged in artisanry and commerce.

One community participated actively in the project – taking the initiative in requesting seedlings, transporting them from the nursery and planting. Community residents also cared for the seedlings and trees by clearing, for example, weeds and undergrowth in the summer to prevent forest fires. This particular community was well aware of the problems caused by deforestation and the importance of conservation. Moreover, unlike many of the more accessible communities located nearer the urban centres, it had not participated in reforestation projects previously and had yet to experience the frustrations felt by many of those involved in projects. Such frustrations often led to considerable scepticism which posed a major obstacle to community participation.

The project assumed an intermittent character due to the fact that funding had to be approved each year and delays in authorizing budgets often meant that activities were temporarily paralyzed. Experienced personnel had to be dismissed and this affected the subsequent execution of the project. The project, for example, only operated during six months in 1990 and by April 1991, (when the fieldwork for the UNRISD case study was terminated) was still awaiting the financial go-ahead for that year.

The agroforestry project supported by CARE had been functioning for 12 years. The project also received support from the United States Peace Corps which provided promoters, petrol and vehicle maintenance. The number of tree nurseries, producing mainly pine and citrus trees, had increased from 35 to 58.

The project had had to overcome four major problems or limitations. First, at the outset it had concentrated exclusively on reforestation activities. It was realised, however, that land conservation and environmental protection required a broader approach. The scope of the project was, therefore, widened to include other conservation activities such as the construction of terraces on agricultural land and the production and use of organic fertilizer.

Second, the participative element in the project was strengthened so that local residents would decide what species to plant and where they would be planted. Work and training were carried out mainly in groups although some individual training was also provided. Initially, the project focused on planting coniferous trees but, in response to local demands, the planting of fruit trees and vegetables became an important aspect of the project.

Third, problems associated with the use of incentives also had to be overcome. Until 1990, the project relied heavily on food-for-work incentives. For terracing a given area, for example, each person received a specified amount of beans, maize, vegetable oil, rice, and ground wheat. This incentive, however, soon came to be regarded as a wage-in-kind. Participation in the project was dependent on the availability of food. Those carrying out the work would calculate precisely how much food they had earned and if these amounts were not forthcoming tensions arose.

Since 1990, there has been an attempt to transfer responsibilities to the local community authorities and to gradually phase out the food-for-work incentives over a three-year period. Other types of incentives have also been developed, including the provision of vegetable seeds, training courses and group visits to see other projects. The change in the type of incentives used has apparently been fairly successful. While food stocks last, food is provided on an occasional basis only, on the understanding that such incentives no longer constitute a regular feature of the project.

Finally, the project has had to learn to respect local customs and beliefs and to engage in constant dialogue with community leaders and residents in order to avoid misunderstandings. Past experience has shown that differences of opinion between project personnel and local residents can easily be blown out of proportion and lead to major tensions which threaten the project. Project officers have also had to learn never to offer what cannot be delivered.

One of the main projects being implemented in Totonicapan since 1985 is the Communal Forest for Fuelwood Project, supported by USAID. This project has sought to break with the traditional method of encouraging reforestation in Indian communities whereby tree nurseries would be located in the cities and personnel from outside institutions (such as students from the University of San Carlos) would participate in reforestation programmes and receive payment from USAID.

The project encourages reforestation and the sustainable management of communal forests. Particular effort has gone into encouraging communities to thin their forests to provide forest products, to create conditions less conducive to the spread of pine borer disease, and facilitate natural forest regeneration.

A range of materials and incentives were initially provided by the project. In addition to technical assistance, inputs such as seeds, wire, bags and fertilizer were provided free of charge. Tools and equipment for planting were provided on a soft loan basis. To motivate community members, the project provided a number of improved fuelwood stoves. Salaries were also paid to community residents who worked on the project.

The payment of salaries, however, soon led to problems. Other community members resented the fact that only a few people were able to participate in the project and receive payment. Intra-community and community/project tensions intensified. Fences protecting reforested areas were cut, sheep were let in to graze, and forest fires were sometimes lit out of spite. As a result of these problems the emphasis switched away from material incentives towards raising awareness of the advantages of reforestation and sustainable forest management.

Project personnel have had to undertake the laborious task of gaining the trust of Indian communities which, historically, felt threatened by *ladino* institutions. During the 1960s, the forestry division of the Ministry of Agriculture appointed forest inspectors in the different regions of the country. These authorities enjoyed considerable powers. A number of people interviewed in Totonicapan recalled how inspectors imprisoned people, confiscated equipment, accepted and expected bribes. Little changed following 1974 when INAFOR was created. The agency was granted the power to impose fines and corruption was rife. DIGEBOS was created in 1988, apparently with the purpose of eradicating corruption.[3] The agency no longer concerned itself with many of the tasks of regulation and control which were transferred to the national and rural police. While there is a new *esprit de corps* within DIGEBOS, it has been difficult for the agency to develop new relations with Indian communities and overcome suspicion and mistrust.

Project personnel have had to learn to respect the rigid rules of protocol and hierarchy which characterize cultural and political relations in Indian communities, gaining first the confidence of community leaders, without which it is almost impossible to work effectively with the community at large. There have been occasions which have facilitated this process – when, for example, project personnel have supported the Indian communities in certain conflicts over land rights and the use of forest resources. This occurred during the course of this study when one of the project leaders organized a meeting of the leaders of various Indian communities with the governor of Totonicapán to discuss the problem of illegal bark stripping and tree felling which was affecting the communal forests.

As noted in Chapter 5, the project itself had been directly affected by this problem. A number of areas where bark was being stripped had previously been reforested with white pine. Moreover, the efforts of DIGEBOS to encourage the communities to thin forest areas backfired when it became apparent that thinning had merely facilitated the access of the bark strippers into what were previously dense forest areas. As a result, several communities began to lose interest in the type of reforestation and forest management practices promoted by the project.

THE HONDURAN SOCIAL FORESTRY SYSTEM[4]

The Honduran Forestry Development Corporation (COHDEFOR) was established in 1974 with the mandate of regulating the exploitation of the country's forest resources, encouraging more environmentally appropriate agricultural land-use patterns, implementing reforestation schemes, promoting other forest and watershed protection programmes, and establishing industrial processing activities which would add value to wood. Via the nationalization of the lumber export trade, the government also attempted to ensure that foreign exchange derived from the export of timber remained in the country and contributed to national development in general and agrarian reform in particular.

A crucial element in the drive to regulate the use of forest land and promote sustainable forms of forest exploitation was the Social Forestry System

(SFS). The latter sought to incorporate the rural population living in or near forest lands in the management and exploitation of forest resources as a means of protecting the forests by reducing the incidence of forest fires, illegal tree felling, cattle grazing in forest areas, and clear cutting associated with shifting agriculture and the expansion of pasture land. For this approach to be successful it was intended that the local population obtain tangible benefits from the exploitation of forest resources.

The SFS involved organizing peasants in agroforestry co-operatives engaged mainly in the extraction of pine resin. A number of co-operatives were also set up to incorporate members of Indian tribes. In addition, the SFS ensured that tribal communities received much higher stumpage fees from the lumber companies than had previously been the case. The SFS was to provide the co-operatives with credit and equipment, as well as technical assistance and training both in productive activities and co-operative organization. Other projects associated with basic needs provisioning and the promotion of silvi-pastoral, agricultural or agroforestry practices were also implemented. Approximately 125 co-operatives were set up during the mid-1970s, and between 1974 and 1977 received support from COHDEFOR. Each co-operative was assigned, via an annual contract, a specific area of forest to exploit and manage. The incidence of forest fires in these areas fell sharply.

The SFS, however, soon encountered serious problems and by the early 1980s it was more of a statistic than an effective programme. Relations between the agroforestry co-operatives and COHDEFOR quickly deteriorated and many co-operative members came to feel highly exploited by the resin and lumber processing companies, and used by COHDEFOR simply as instruments to protect pine forest areas.

As explained in one document presented at the first national conference of the co-operatives, held in 1977:

> It is not true that Decree 103 [which established COHDEFOR and the SFS] benefits the peasant integrated in the agroforestry co-operatives. On the contrary, he is merely relegated to the status of an unpaid forest guard. Moreover, other problems have arisen such as excessive interference of the state in our internal affairs, being forced to comply with onerous contracts, while COHDEFOR extends permits to [lumber] companies in the areas where we operate and where pine resin can still be extracted.

The number of participants in the SFS declined and income levels from co-operative activities fell. In 1986, 104 co-operatives with 4,080 members continued to operate in eight of the country's nine departments. By the early 1990s, approximately 80 groups were operating.[5]

In the latter half of the 1980s, the SFS entered a new phase with the development of 50 Areas of Integrated Management (Areas de Manejo Integrado – AMI), each of between 1,000 ha and 10,000 ha (Synnott, 1989:108). With the support of the FAO, COHDEFOR sought to incorporate local communities in a system combining agriculture with the sustainable use of forest resources. The participants in this programme, who numbered 2,500 in 1987, worked in extractive activities organized by the lumber companies. In 1987, the AMI covered some 400,000 ha of forest (90 per cent pine and 10 per cent broadleaf). The number of AMI was expected to increase in later years (SECPLAN *et al.*, 1990:150).

The impact of the SFS in arresting the process of forest degradation and improving the levels of living of local populations has been minimal. Moreover, as indicated below, the programme's impact on the lives of local populations in certain areas has been somewhat contrary to the letter and spirit of the law.

Technical reasons partly account for the demise of the SFS. Shortly after the programme was established the price of pine resin, one of the principal products of the agroforestry co-operatives, fell. The price problem, however, also related to the failure of a COHDEFOR company to effectively regulate prices. The company collapsed in the late 1970s, partly as a result of corruption and maladministration. As a result, the co-operatives had to accept the prices imposed by a monopolistic market where just two companies controlled over 80 per cent of the market. While significant profit margins were obtained by the four companies which purchased the pine resin, the co-operatives were paid very little for the resin. It has been estimated that the total income of the agroforestry co-operatives amounted to 3.3 million lempiras, the equivalent of 809 lempiras per member per annum. This was the equivalent of 67 lempiras a month or one-third of the minimum wage.

Other constraints associated with the limited administrative experience of the co-operatives and capacity of COHDEFOR to provide technical assistance, equipment, inputs (credit-in-kind), and other support services also affected the operations and profitability of the co-operatives (COHDEFOR, 1988).

To understand the failure of the SFS, however, it is not enough to consider technical, administrative and pricing problems. It is also necessary to refer to a very different set of conditions involving social and political aspects associated with a) changes in the correlation of social forces at the national and international levels, and b) the reality of social structure and forms of organization at the local level where projects and programmes were implemented.

The starting point for this enquiry is the specific configuration of socio-political conditions which gave rise to the progressive legislation of 1974 that ushered in the SFS and the agrarian reform programme. At the beginning of the 1970s, social tensions in many rural areas had reached breaking point. Extreme poverty and a highly inequitable system of land distribution were fairly generalized throughout the country. Moreover, it was apparent that the exploitation of the country's extensive forest resources by mainly foreign-owned lumber companies, was contributing very little to national development in general and the rural economy in particular.

The level of peasant organization increased dramatically during this period. In several regions, the Catholic church sided with the rural poor and called for agrarian reform and measures to regulate the exploitation of forest resources. This level of mass mobilization and organization was not simply a response to highly inequitable structures of distribution of wealth and income that had existed for decades. Levels of peasant organization had increased during the

1960s through the process of 'peasantization' of both Honduran banana workers, many of whom had union experience, and of Salvadorean migrant workers often from urban areas. It also resulted from the support of groups or agencies other than the Catholic church which included, for example, the Christian Democratic Movement and left-wing political groups. The student movement was also extremely active during this period.

The repressive character of the military government had been neutralized during this period, partly in response to the overwhelming popular support for the military which arose during the 1969 war with El Salvador. Following the war, the army adopted a more nationalistic and progressive stance. A group of younger officials, motivated to some extent by the experience of military reformism in Peru and Panama, gained considerable influence. In December 1972, this reformist sector assumed power through a military coup and introduced programmes of agrarian and forestry reform.

The correlation of social forces which had given rise to this reformist approach was soon, however, to alter dramatically. A powerful anti-reform movement, led by the large lumber companies, emerged during the latter half of the 1970s. The media was saturated with information and commentaries denouncing 'communist' measures and calling for privatization, particularly of the lucrative lumber export trade.

The principal organization opposing the forestry reform was the Honduran Lumber Dealers Association (AMADHO) which formed part of the country's most powerful non-governmental organization, COHEP, representing the private business sector. Two other organizations linked to lumber interests were also represented in COHEP: the National Association of Industrialists (ANDI); and the National Federation of Honduran Farmers and Ranchers (FENAGH). During the period when opposition to the reform project was most intense, one of the country's largest lumber dealers was both president of FENAGH and of COHEP.

The group of progressive officials within the army gradually lost influence and corruption within the military increased. The excuse to organize a coup to do away with the reformists came in 1975 when the *Wall Street Journal* reported that a transnational company with extensive banana interests in Honduras had bribed the Minister of the Economy to the tune of 1.25 million dollars. The Honduran president, Lopez Arellano was also implicated in the affair which cost the reformists their short period in office.

Reformist policies gradually waned during the latter half of the 1970s and were dealt a more decisive blow in 1982 when the civilian government of Suazo Cordova replaced the agrarian reform programme with a land titling scheme and began the process of reprivatization of companies which had been taken over by COHDEFOR. Controls on the allocation of areas for forest exploitation were also relaxed.

Pressure brought to bear by the lumber companies had been reinforced by the United States government which made its views very explicit shortly after the Liberal Party had won the 1981 elections. The United States ambassador in Honduras formally announced a series of policy guidelines which it expected the Honduran government to follow if United States aid was to be increased. According to these guidelines, government policy should encourage land titling instead of land redistribution, halt further nationalization of private enterprises, reprivatize state companies, particularly those pertaining to COHDEFOR, and increase stumpage payments to COHDEFOR in order to encourage reforestation programmes.

During the first half of the 1980s, the state retained control of the most lucrative link in the forestry sub-system, namely, the export trade in lumber products. Following elections in 1985, however, the new government toed the neo-liberal line and reprivatized the export trade. During this period individuals representing powerful lumber interests gained considerable influence in policymaking circles. In 1983, AMADHO had pressured the government to change the governing board of COHDEFOR and had two of its members appointed. In 1985, the owner of one of the country's largest lumber companies was proposed as presidential candidate by a right-wing faction of the National Party. The head of AMADHO was a member of the National Congress and an influential member of a commission set up to evaluate COHDEFOR.

The new government also bowed to the demands of the lumber companies which had demanded the right to exploit clearly defined areas of forest land on a more permanent basis. In 1986 COHDEFOR established a system of 'tributary areas' (*áreas tributarias*) which effectively subdivided the country's forest land among the different lumber companies, granting them areas with sufficient wood to enable them to continue production for an indefinite period. Each company was expected to operate according to a management plan approved by COHDEFOR which would ensure the sustainable exploitation of the forest. By 1989, however, only one company had drawn up such a plan (SECPLAN *et al.*, 1990:151).

COHDEFOR's capacity to regulate the use of forest resources seemed set to decline even further in 1990 when municipal authorities were granted the right to administer the use of forest lands in their municipalities. These relatively weak local level authorities generally lacked either the technical/administrative capacity or the power to ensure that large companies with considerable economic and political muscle would abide by regulations governing the sustainable use of forest resources. Moreover, at the local level it was sometimes the case that the lumber companies were in collusion with municipal leaders.

This change in the balance of social forces is crucial for understanding the demise of the SFS and the decline in institutional support for the programme. COHDEFOR ceased issuing contracts which had given the co-operatives the right to exploit a given area. Yet having operated for many years and undertaken forest management practices, the co-operatives naturally felt they had a just claim to these areas and resented the incursion of the lumber companies. Many companies tried to alleviate tensions by contracting the co-operative members as labourers to cut and haul trees.

The experience of the Social Forestry System on the ground

In the space of 16 years, then, there was a dramatic about-turn in the organization of the forestry sector in Honduras. This has had major implications for the local communities which were supposed to benefit from the SFS. To examine in more detail the operation of the SFS on the ground, we turn now to a case study of agroforestry co-operatives and Indian tribes in the department of Yoro.

Located in the mountainous region of north-western Honduras (see Map i.1), Yoro has a population of approximately 320,000 and contains some 758,000 ha of both pine and broadleaf forest. Forest resources in the department have been subject to rapid deforestation during the past three decades, mainly due to the operations of lumber companies. Extraction and lumber processing techniques have been highly inefficient in terms of natural resource use. Lumber companies generally operate with out-dated machinery using circular saws.

During the early 1970s, 19 of the country's 149 lumber companies operated in Yoro. Today 11 lumber companies continue to operate in the department, producing, in 1990, 112 thousand cubic metres of wood, principally pine. Competition for raw material supplies has increased. Nearly half of the department's timber production is produced by just one company – Yodeco – which had been nationalized during the 1970s and reprivatized in 1986. Constraints on raw material supplies in the department are such, however, that Yodeco must obtain the bulk of its logs from the neighbouring department of Olancho, encroaching on forest areas formally assigned to another lumber company.

The operations of the lumber companies, combined with the effects of forest fires (mainly in pine forest areas), and shifting agriculture and cattle expansion (mainly in broadleaf forest areas) have led to serious environmental destruction in Yoro.

Many areas which were once forest land are now largely unproductive scrublands. Erosion is serious and flooding has become more commonplace in lowland areas. Water levels of many streams and rivers have declined significantly, particularly during the summer months. A local peasant recalled vividly how nature had changed during his lifetime:

> You see along that ridge where the coffee trees end – 25 years ago the forest was impenetrable. When we entered the forest we would have to cut paths with our machetes and we bounced along on a carpet of leaves. We never saw the sun. Many locals would enjoy trying their hand with a sharp axe to fell trees a metre thick. The streams were crystal clear and when you washed your hands in the water the cold would penetrate to the bone. Today all we see around us is land cleared by peasants and the chainsaw of the lumber company. Fuelwood is now becoming scarce and some streams have dried up completely.

The agroforestry co-operatives of Yoro

Three agroforestry co-operatives operated in Yoro. The largest was the Regional Agroforestry Co-operative of Yoro (CAFREYOL) which, until 1990, comprised 14 peasant groups, totalling 300 members engaged in a variety of activities. Two other co-operatives, 'Altos Pinares' (24 members) and '20 de Marzo' (62 members) both extracted pine resin.

The livelihood alternatives of the local population in the areas where these co-operatives exist have, traditionally, been extremely limited. A few people would find work in the local sawmills but wages were very low. It is estimated that 70 per cent of the department's sawmill workers earn the minimum daily wage which in 1991, amounted to 8.64 lempiras ($1.63). It was common for people to migrate to other areas in search for work.

Those who joined the co-operatives were generally poor peasants who produced maize, beans and yucca on small plots of land, usually of no more than a third of a hectare. Soils in the area are generally unsuited to agriculture given the steep terrain, rocky conditions and the highly acidic nature of pine forest soils. Maize yields of just 20 qq (909 kg) per ha and bean yields of 10 qq (455 kg) were the norm. In contrast, farmers producing in the fertile Aguán Valley some 30 km away could obtain yields three times greater.

The co-operatives were established during the mid-1970s under the auspices of the SFS. Their productive lives, however, were generally short-lived. The 'Altos Pinares' and '20 de Marzo' co-operatives began to extract resin in 1974 but ceased activities three years later partly as a result of the type of problems outlined earlier regarding low prices and lack of technical support from COHDEFOR, but also due to conflicts which arose when the forestry corporation granted a local lumber company the right to exploit the same areas where the co-operatives were supposed to operate. Faced with limited alternative employment opportunities and an upswing in the price of pine resin, the old membership regrouped in 1989 and recommenced activities. COHDEFOR has since provided some support mainly in the form of training in techniques of resin extraction and forest protection practices, as well as financing for materials and tools.

The co-operative members have been unable, however, to earn a living income. The average monthly income of members of the two co-operatives in 1990 was 74 ($15) and 88 lempiras ($18) respectively, approximately half the minimum wage. The members of these co-operatives have been unable to break out of the poor peasant mould. Not only do they continue to farm their subsistence plots but many continue to migrate in search of work as agricultural labourers for several months of the years.

Several of the peasant groups which form CAFREYOL also commenced activities during the mid-1970s. They too engaged in pine resin extraction but could not make a living from this activity. They went into a long period of recession with members struggling to survive, cultivating their subsistence plots and finding about two weeks work a month as wage labourers, and earning little more than 20 dollars a month. It was not until the latter half of the 1980s, when CAFREYOL diversified its activities and decided to take a stand against COHDEFOR and the lumber companies that the socio-economic situation of the members improved. This experience is reviewed in the following chapter.

The Social Forestry System and the Xicaque Indians[6]

Approximately 18,000 people live in the tribal communities of the Xicaque or Tolupan Indians, grouped in some 28 tribes scattered throughout Yoro or adjacent areas in other departments. Unlike the Mayan Indians, the Xicaques were not agriculturalists but were semi-nomadic hunters and gatherers, living mainly along the Atlantic coastal strip. With the arrival of the Spaniards they fled towards the isolated mountain regions of the interior. In the forests of Yoro they lived off game, root plants and other products such as honey. In 1864, 21 tribes received title to their lands. There is, however, considerable confusion over boundaries.

During the past century, the Xicaques came to depend increasingly on agriculture for subsistence provisioning. This was partly as a result of restrictions on movement (necessary for hunting and gathering) which resulted both from the colonization of the area by *ladinos* who staked claims to land, and the Xicaque practice of incorporating poor *ladino* peasants into the tribal structure. This was achieved in two ways: through marriage with Indian women or simply by granting permission to *ladino* peasants to live on and farm tribal lands. Those who married Indian women became known as 'incorporated *ladinos*' (*ladinos incorporados*) and obtained usufruct rights which were granted as long as the *ladino* remained married to an Indian wife. The children of the marriage obtained Indian status. The other 'resident *ladinos*' (*ladinos avecindados*) could farm the land and exploit the forest but neither they nor their children had rights to the land. They could remain only as long as was deemed appropriate by the tribal authorities. This system of incorporation functioned as a sort of collective survival strategy. It served, in particular, a twofold purpose: to reduce racial tensions; and to secure claims to tribal lands by populating areas under potential threat of encroachment by landowners and logging companies.

Agriculture remained, however, extremely rudimentary and as integration into the market economy increased, notably during the 1960s and 1970s, many tribes came to depend on the sale of trees to obtain the cash income they needed.

Until 1974, the lumber companies that wanted to extract timber from a tribal area, negotiated directly with the tribal chiefs (*caciques*). A derisory stumpage fee of 2 lempiras ($1) was paid per cubic metre. Often, though, the tribes remained unpaid, seemingly as a result of corruption within the Honduran Indigenous Institute which acted as an intermediary between the companies and the tribes. By the mid-1970s, a debt of 30,000 lempiras ($15,000) had accumulated. Often the stumpage was appropriated by the *cacique*. As a result, intra-tribal tensions intensified and there were numerous instances of tribal chiefs being forced to resign, imprisoned or physically injured as a result of disputes.

The new forestry legislation of 1974 introduced several important changes in the traditional system. COHDEFOR intervened as an intermediary between the lumber companies and the tribes. The stumpage fee was increased to 12 lempiras ($6) per cubic metre. It was intended that this money would benefit the tribe as a whole and not just the *cacique*. It was also intended that the Indian population would take a more active role in activities associated with forest protection.

Many *caciques* resented these measures which they felt undermined their authority. Some entered into illegal transactions with the lumber companies. Corruption involving forestry officials was widespread. Certain officials would accept bribes from the lumber companies and would pressure the chiefs to negotiate with the companies. It appears there were considerable tensions between many officials based in Yoro and the forestry corporation. Being assigned work in an isolated department was considered more of a punishment than a job. Not only were COHDEFOR employees poorly paid but pay cheques often arrived late. Feeling alienated by the corporation, many technicians had no qualms about entering into deals with lumber companies and undermining the letter and spirit of the SFS by pressuring *caciques* to supply the companies with the wood they required.

Indian chiefs who did not toe the line were often forced to resign or found themselves confronted by a powerful array of forces when it came to elections. Local mayors and leaders of the major political parties (Liberal and National), military officers, landowners, lumber companies and forestry officials disposed of numerous means to ensure the acquiescence of many *caciques*. In this milieu, the institution of *cacicazgo* lost much of its legitimacy, particularly among the young and those of mixed race (the *indio-ladinos* from mixed marriages). Intra-tribal tensions intensified, several *caciques* and tribal members were assassinated or imprisoned. As tribal institutions weakened several landowners and lumber companies took advantage of the situation and invaded tribal areas. Stumpage payments remained unpaid and by the early 1980s, a debt of half a million lempiras ($250,000) had accumulated.

A number of projects promoted by the SFS failed dismally. COHDEFOR had set up five co-operatives to extract pine resin. The co-operatives, however, ran into debt, both because of falling prices and the fact that they had to purchase the necessary equipment from COHDEFOR. Corruption appears to have been widespread. Considerable confusion exists as to what happened to the payments. While the co-operatives claimed that they had paid the project promoters, COHDEFOR argued that they had received nothing. The tribes tended to see these projects as something forced upon them by COHDEFOR and seriously doubted any good intentions the agency may have had.

The SFS also provided equipment and training for professional bee-keeping activities. Artisanal apiculture had always been practised by many Indians who, traditionally, had cut that part of a tree which contained the hive and located it nearer their homes. The project, however, had the effect of eliminating these traditional practices. Participation in the project soon declined as debts accumulated and the threat of the arrival of the African bee put people off the idea of bee-keeping.

Faced with this situation of corruption, violence and failed projects, a group within the forestry corporation attempted to improve relations with the tribes and drew up a more integrated development strategy. From this initiative there emerged, in 1979, the Indigenous Co-operation Project of Yoro which was supported by a grant of 3 million lempiras ($1.5 million) from the EEC. Working with five tribes, this pilot project sought to promote the type of actions originally proposed by the SFS including forest management, the setting up of small lumber mills, the introduction of silvi-pastoral and agroforestry land-use systems, as well as schemes to resettle dispersed populations in communities, with improved access to social services.

Certain development organizations, such as the Swiss agency CSD, tended to exclude the Indian population from their programmes and worked more with communities of *ladino* peasants. Tensions also arose from USAID's programmes which attempted to involve the Honduran Indigenous Institute. The mistrust which existed between many tribes and the Institute intensified when the activists participating in the programme received generous payments for field expenses. The USAID programme in the area was eventually disbanded although Peace Corps volunteers continued to operate.

Several important conclusions or lessons can be drawn from this review of the experience of the Social Forestry System in Honduras. First, when trying to identify the principal problems which affect the implementation of conservation projects and programmes it is important to look beyond the 'technical' constraints associated with budgets, human resources, prices and administrative questions. Such limitations are often symptomatic of broader changes occurring in state priorities which relate to political economy questions. What the Honduran case study reveals clearly is that such changes occurred within a specific socio-political context. A major shift had taken place in the balance of social forces which resulted in fundamental changes in government policy that favoured the lumber companies and marginalized the agroforestry co-operatives.

Second, the case study of the Xicaque Indians illustrates that progressive policies designed at the central level can be easily undermined at the local level if they are superimposed on inegalitarian local level social structures and if they do not have the support of local authorities and dominant interest groups.

Third, the Honduran experience raises questions regarding the sustainability of what might be called progressive 'eco-social' initiatives (that is, initiatives promoting environmental protection or rehabilitation and social development benefiting low-income groups) when these are promoted by the state and are not reinforced by active grassroots mobilization and support. To the extent that the general social orientation of state policy (who gains and who loses) reflects the balance of social forces and that the latter is in a state of flux, significant changes in the orientation of state policy can be expected, particularly when progressive policies are involved which threaten powerful vested interests.

Finally, the analysis of how political economy issues can affect the design and implementation of environmental protection initiatives must occur at different levels, taking into account: the socio-political character of local level structures and institutions; the role of interest groups at sub-national levels (for example, municipal authorities, decentralized structures of government agencies, logging companies, etc.); and the way in which decision-making processes at the central state level are shaped by pressures and influences involving both national and international actors and organizations.

Chapter 11

Alternative approaches: Revolutionary change and grassroots mobilization

The mainstream forest protection and tree planting initiatives reviewed in previous chapters generally attempt to promote adjustments in socio-economic and technological systems which do not seriously question or transform established social and political structures. Very often it is the traditional or modern élites – for example, loggers, large commercial farmers, tribal chiefs, transnational corporations, eco-tourism companies, urban middle classes and the scientific community – which benefit from such schemes. The rights, needs and cultural perspectives of many peasant and indigenous groups are often ignored. When they are addressed, the quantity of resources which actually flows in their direction is usually so limited as to make any significant progress along a sustainable development path virtually impossible.

Other approaches to forest protection exist, however, which attempt to change established structures of resource distribution and power relations. In the context of Central America, two such approaches seem particularly relevant. The first relates to the process of revolutionary change which occurred in Nicaragua following the overthrow of the Somoza dictatorship by the Sandinistas in 1979. The second concerns various instances of grassroots mobilization which have occurred periodically throughout the region. The potential and limitations of these two alternative approaches to forest protection are considered briefly below.

RADICAL STRUCTURAL CHANGE: AGRARIAN REFORM AND HUMAN RESETTLEMENT IN RÍO SAN JUAN, NICARAGUA

The analysis of the causes of deforestation presented in Chapter 2 indicated that migratory agriculture and the colonization of agrarian frontier areas by land-seeking peasants constitute two of the fundamental processes underpinning deforestation. Such processes are intimately tied up with questions of limited employment opportunities outside of agriculture, land concentration, and the highly restricted access of peasant producers and landless labourers to land, basic inputs, consumer goods and support services. In short, processes of deforestation are part and parcel of a particular style of development characterized by highly skewed patterns of resource distribution. When addressing the question of how best to protect forest resources, it is pertinent, therefore, to consider policies and programmes which rapidly transform structures of resource distribution with the purpose of providing disadvantaged groups in rural areas with improved livelihood opportunities.

To examine these aspects, a case study was undertaken of Nicaragua's southern-interior department of Río San Juan where intense processes of population resettlement and agrarian reform had taken place following the Sandinista revolution.[1] These had been motivated by a combination of concerns. The military was eager to depopulate remote areas in order to create free-fire zones and undermine the social base of support of the *Contra* rebel forces who were attempting to overthrow the Sandinista government. Planners and policymakers concerned with the question of economic and social development considered it important to provide scattered populations in such areas with support services and land which was more suitable for agricultural, notably food, production. A key purpose of the study was to ascertain the degree to which this alternative approach contributed to sustainable development, taking into account both environmental and social welfare criteria.

By the end of 1983, some 1,400 families, approximately 70 per cent of the population that lived in the interior of the department, had been evacuated. The majority were relocated in 16 settlements situated in more accessible areas on the western side of the department, nearer lake Nicaragua and the main road linking the departmental capital, San Carlos, with the rest of the country. Men, and some women, of working age were generally organized in co-operatives which enjoyed access to land, credit, and inputs, as well as social services, notably health and education.

Throughout the 1980s, major changes occurred in agrarian structure in Río San Juan. The percentage share of land occupied by peasant producers and co-operatives increased from 13 per cent to 44 per cent between 1978 and 1989. State farms, which were non-existent before the revolution, made up half the total farm area while the area occupied by large private producers declined from 87 per cent to 6 per cent over the same period. Eighty-nine co-operatives were established with a total membership of 1,500 members (15 per cent of the department's working population). In all, some 2,475 families (approximately one-third of all families in the department) received land titles.

From an environmental point of view, the resettlement and agrarian reform processes yielded extremely mixed results. While the rate of deforestation in the interior of the department declined considerably, deforestation increased dramatically in resettled areas. Estimates by development specialists in the area indicate that the forested area in zones which came under the agrarian reform programme declined by roughly 50 per cent between 1985 and 1991.

The reasons for this high rate of forest destruction have to do with the type of development model promoted by the Sandinista government and the response of peasant/co-operative producers to the economic crisis which beset the country from the mid-1980s onwards. While the process of structural transformation which occurred in Río San Juan formed part of a broader development strategy that recognized the need to restrict agriculture and expand forestry in the interior regions of the department, there was little attempt to implement an 'agroforestry reform' in the sense of cultivating new attitudes, technologies and land-use practices conducive to the preservation or sustainable management of forests and the incorporation of trees in farming systems. Technical assistance and training programmes focused almost exclusively on the use of technological packages associated with modern agriculture, and administrative and socio-political aspects of co-operative organization, rather than the promotion of farming practices more appropriate for the type of agro-ecological and socio-economic conditions prevailing in the area. In the context of a war which threatened to cut supply routes with the rest of the country, the goal of achieving departmental food security or self-sufficiency received high priority. As a result, land clearance for basic grain production proceeded apace in resettled areas.

Moreover, people remained highly dependent on the use of forest products. Wood was required to build over a thousand homes in the settlements. As the economic crisis intensified during the latter half of the 1980s, some families turned to the commercial exploitation of forest products (notably fuelwood) as a means of obtaining supplementary income. In the area where resettlement took place, there was a marked increase in the fuelwood trade. It was common for those trading in fuelwood to spend three days amassing 1,000 pieces which would be sold in 10 bundles or *cargas*. In 1990, each *carga* fetched $1.25. In a month, a family might be able to obtain around $50 from the sale of fuelwood, roughly the equivalent of an agricultural labourer's monthly wage.

The need to find additional sources of income was further reinforced during the late 1980s when the Sandinista government introduced economic stabilization policies which reduced or eliminated many of the subsidies which the co-operatives had enjoyed. Technological packages which were highly dependent on imported agrochemicals, machinery and fuel were no longer viable. Such problems intensified during the early 1990s when, following the change in government, much tighter controls were introduced on credit and many support services crumbled. These changes in economic policy affected yields and encouraged families to engage once again in shifting agriculture.

From a social point of view, the results of this alternative development approach were extremely mixed. Some important gains in living levels were achieved early on. The economic and social situation of many families, however, deteriorated sharply during the latter half of the 1980s and early 1990s as the effects of war, economic crisis and stabilization policies reduced the availability of basic goods and services and cut subsidies.

A considerable number of families, or certain members thereof, began drifting back to agrarian frontier regions. The prioritization of the co-operative sector in social and economic assistance programmes meant that many private peasant producers had remained marginalized in the structure of resource allocation and had had to endure extremely poor living conditions. Several co-operatives disbanded and sold assets, and co-operative members returned to the interior. While recolonization resulted largely from the economic difficulties experienced by the co-operatives, it was also due partly to a cultural rejection of certain forms of co-operative organization which were imposed from above.

During the early 1990s, the flow of people returning to agrarian frontier regions increased as ex-*Contra* rebels, 'returnees' from Costa Rica and migrants from other parts of the country moved into these areas, intent on recuperating old plots or staking new claims. This recolonization process posed a serious threat both to forest resources and the reserve status of a large protected area in the south-eastern corner of Nicaragua.

This process of return, however, was very uneven. There are indications that several co-operatives that were more consolidated in organizational, administrative and material terms remained intact. There were also signs that many who returned did so for relatively short periods, alone or accompanied by friends or just one family member, instead of the entire family.

Throughout the past decade many women and youths had come to appreciate the benefits of living in areas accessible to social services and alternative employment opportunities in urban centres and were unwilling to return to the harsh conditions of life in the interior. Moreover, women who were widows or heads of single-parent households sometimes joined co-operatives and received land titles. It seems that certain families were now keeping a foot in both camps, so to speak: retaining their home base in the settlements and using land obtained through the agrarian reform for cattle grazing, while one or two male members returned for short periods to the interior to clear land for grain production.

GRASSROOTS MOBILIZATION

The types of conservation and development approaches reviewed above generally have in common the fact that they constitute initiatives that were inspired and co-ordinated by agencies external to the areas and communities directly affected by deforestation. Some of the most effective initiatives in the field of forest protection, however, have arisen from the organized efforts of local populations seeking to defend their natural resource base, either confronting directly – through various forms of protest and struggle – the activities of private and government entities promoting deforestation, or exerting pressure through institutional channels to prompt changes in government policies and programmes.

Such initiatives fall within the comprehensive definition of 'participation' referred to in Chapter 9 which implies both 'the willing, informed and active involvement of people in the decision-making process on issues affecting their lives' and 'organized group action and the sharing of political and managerial power' by hitherto disadvantaged groups (UNRISD, 1979; see also Barraclough, 1991). The discussion and literature on participation within the region, however, often overlook this latter aspect and focus primarily on the question of consultation and dialogue during the project cycle or planning process.

Some of the Central American experiences in this field of 'popular mobilization' are already well known. The case of the Kuna Indians in Panama, for example, has been written about extensively (Archibold, 1992; Houseal *et al.*, 1985; Myers, 1990). The Kuna were granted a high degree of autonomy and full legal title to their land following a long history of conflict with the government which included armed revolt in the mid-1920s. Since the mid-1970s, the Kuna have undertaken several initiatives to protect their 3,206 square km reserve from encroachment by outsiders. Underpinning the colonization of tropical forest areas was a government policy that promoted the settlement of 'land which is not being exploited'. Following the commencement of road construction in 1970 to link the interior of the country to the Caribbean coast, many settlers arrived from other parts of Panama to clear the forests bordering Kuna territory or actually within the reserve.

To arrest this situation a number of Kuna decided to settle the area during the latter half of the 1970s, in order to demonstrate that it was being exploited. After a few years, however, it became apparent that the soils were not appropriate for agriculture. On receiving advice from CATIE and support from several international conservation agencies, the Kuna set up, during the early 1980s, a wildland park and botanical sanctuary in an area covering 600 square km. In 1987, the Kuna General Congress approved a proposal that the territory be managed as a biosphere reserve with different management zones, determined on the basis of ecological characteristics (Archibold, 1992: 30). As pointed out by Myers:

> ... the Kunas are perhaps the only indigenous group in tropical areas of developing countries who systematically administer their own natural resources on a large scale. (ibid., 1990:83 – author's translation from Spanish)

Grassroots action to protect the local environment and culture can, of course, assume diverse forms. As in the case of the Kuna Indians, indigenous groups have occasionally resorted to armed struggle. During the 1980s, several Miskito Indian organizations in Nicaragua took up arms against the Sandinista government. Such a response was instrumental in prompting a radical rethink of government strategy regarding economic and social development in the Atlantic coastal regions. It also put the issue of regional autonomy firmly on the government's agenda. The question of how to exploit, and who should exploit, the region's extensive forest and other natural resources was central to the change in strategy.

National governments have often responded to both violent and peaceful forms of protest with the use of force. Perhaps the worst instance of repression occurred in 1932 in El Salvador. Indians wanting to maintain communal lands and resist the encroachment of coffee plantations joined the uprising led by Farabundo Marti and occupied several towns in the country's coffee-growing region (Burger, 1987:74). The army responded by massacring 30,000 Indians and peasants.

Unfortunately, such extreme responses are not a thing of the distant past. In 1978, 400 Guatemalan Kekchi Indians were murdered or injured when fired upon while protesting in a town square at attempts by large landowners to evict them from lands they had farmed for decades but for which they had no title (ibid.:83). The indiscriminate shooting of Indian people by the military has continued to the present day. In December 1990, Guatemalan newspapers reported that the army had killed 13 Indians and wounded several others during a 'peaceful protest' in the town of Santiago de Atitlán.

Protest may assume less obvious forms. A major problem in Honduras, for example, has been the high incidence of illegal forest fires which according to some observers are 'set out of spite because the *campesino* believed COHDEFOR (the Honduran Forestry Development Corporation) has usurped his forest' (Campanella *et al.*, 1982).

The expansion of civil society in Central America in recent years has seen the emergence of organizations representing small farmers and Indian groups in several countries. In various instances this has facilitated mobilization and also meant that pressure can be brought to bear via institutionalized political channels. The importance of such efforts is borne out by the experience of one of the region's show-pieces in community-based reforestation – Hojancha, Costa Rica (de Camino, 1989; Current, D., 1991). Here a community organization exerted pressure on government authorities to obtain the types of financial incentives for reforestation which previously had benefited only larger landowners. In 1988, the Department of Peasant Forestry Development (DECAFOR) of the Costa Rican Forestry Service called on the government to modify the incentive scheme. While large landowners opposed these proposals, they were eventually accepted (Current, D., 1991:29–30).

In Honduras, numerous Indian organizations were formed during the late 1970s and 1980s. They were successful in getting the state to formulate the Law for the Protection and Development of Ethnic Groups of Honduras which was intended both to protect indigenous populations and regulate the use of natural resources in areas where such populations were located (SECPLAN *et al.*, 1990:74). Indigenous organizations in Honduras, however, have often been short-lived or rendered unrepresentative and ineffective as a result of fragmentation, co-optation and corruption. Many organizations experienced considerable instability. A Pan-Indian movement was established in 1977 to defend Indian lands. Control of the organization, however, was largely in the hands of *ladinos* and it was soon disbanded (SECPLAN *et al.*, 1990:73). Prior to 1985, the organizational efforts of the Xicaque Indians in Yoro experienced similar difficulties.

The capacity of Honduran indigenous organizations to operate as effective pressure groups has, then, remained fairly limited. A USAID report on the environment in Honduras noted that despite the fact that the rights of Indian groups to maintain their cultural forms and land base were recognized in law, it was clear, during the early 1980s, that 'they have little serious support from the National Agrarian Institute' (Campanella *et al.*, 1982). During the mid-1980s, both the Xicaques and the Pech Indians formed what appear to be stronger organizations (ibid.).

In a number of countries, groups affected by deforestation and encroachment are finding it easier to articulate and publicize their grievances because of alliances with urban-based ecology groups which tend to have greater access to the media, certain state agencies and policy decision-making circles. This is apparent, for example, in Costa Rica, El Salvador and Nicaragua.

In the course of this study, several other cases were encountered where grassroots organization and mobilization yielded positive effects in both environmental and social terms. It should be stressed that such initiatives rarely seek to halt the exploitation of forests. Rather they attempt to ensure that the benefits of forest exploitation accrue primarily to populations living in the area and that the natural resource base on which that population depends will still be there for the benefit of future generations.

Grassroots initiatives in Yoro, Honduras

In the preceding chapter, we analysed the experience of the Social Forestry System in Honduras which had provided few tangible benefits for the target population in the department of Yoro. Rather, local élites and corrupt officials had succeeded in undermining implementation and appropriating benefits for themselves. Here we take up the story of certain peasant and Indian groups in Yoro to see how grassroots initiatives were instrumental in improving their situation. Certain disadvantaged groups responded by creating new forms of socio-economic or socio-political organization and establishing what, in effect, was an alternative social forestry system.

The regional agroforestry co-operative of Yoro – CAFREYOL[2]

In 1985, there was a qualitative change in the situation of several peasant groups which had participated in the social forestry system when they merged to form CAFREYOL. Having abandoned pine resin extraction and broken the link with the processing companies, the co-operative began cutting and processing wood in an area of approximately 3,000 ha consisting mainly of pine forest. Production increased significantly from 1,249 to 5,650 cubic metres between 1985 and 1990.

The activities of the co-operative, however, were opposed by an array of institutions and interest groups at the departmental level that included local lumber companies, representatives of COHDEFOR, and departmental and municipal authorities, which, in line with national policy, sought to secure forest areas for the lumber companies.

Support for CAFREYOL came from certain NGOs and national trade union organizations. The National Council of Rural Workers (CNTC) provided substantial training and technical assistance in areas associated with co-operative organization, administration, education, production and social affairs. The Swiss development agency, CSD, contributed funding for productive activities and technical training in accordance with development plans elaborated by the co-operative and approved by the funding agency.

The degree of support from such agencies contrasted sharply with that of the state forestry corporation, COHDEFOR. The co-operative came to regard the state agency more as a hindrance than a help. Requests by the co-operative to increase the number of members per group (restricted by

COHDEFOR regulations to 15) and to obtain access to larger areas were repeatedly rejected or ignored by COHDEFOR. Forest areas granted to the co-operative were sometimes those affected by tree disease. The co-operative saw this as part of a general strategy to favour the lumber companies. Since the creation of the system of Integrated Management Areas (AMI), referred to in Chapter 10, the co-operatives have been further marginalized by COHDEFOR. Requests by the co-operatives to fell trees in the Tributary Areas assigned to the lumber companies have usually been rejected while those that are granted often carry with them the condition that the trunks must be sold to the companies.

The purpose of the AMI was to involve the population, living within the limits of the Tributary Areas, in agricultural and tree felling activities. Employment in such activities was expected to reduce the pressures on the peasant population to constantly clear forest for agricultural purposes. In addition they were encouraged to produce subsistence crops in other areas. The AMI programme has attracted the attention of several agencies (including CATIE) and experts, some of whom see it as a potential model to be applied elsewhere (see Poore, 1989:201).[3]

The experience of the AMI in Yoro, however, was less than encouraging. Five such areas had been established, each one having been assigned an average of 2,500 ha of forest. This form of organization generally proved unsuccessful and by 1991, just one of the AMIs remained active.

Areas assigned for crop production often consisted of marginal land on hillsides. Tree felling activities were organized by the company which had been assigned a Tributary Area, in co-ordination with COHDEFOR. The company also set the price at which the trunks were sold. Such a system essentially relegated the people involved to the status of paid company employees.

This approach contrasted sharply with the more integrated development approach promoted by CAFREYOL. The co-operative association attempted to encourage the peasant groups to add value to their products through different types of processing activities and by establishing more direct links with consumers, thereby cutting out the middleman.

Unlike the agroforestry co-operatives which had links with national union federations, the groups involved in the AMI programme remained independent and politically weak. In practice, there was little that was 'integrated' about the Integrated Management Areas. COHDEFOR lacked the multidisciplinary approach or personnel necessary to promote activities associated with social development and training.

The agroforestry co-operatives also came under pressure from other quarters. Local departmental and municipal authorities granted tree felling permits to local community groups who formed part of their political base of support. These groups often undercut the co-operatives by selling at very low prices to the lumber companies.

In the midst of opposition from local lumber companies, the regional offices of COHDEFOR and departmental and municipal authorities, the co-operative did not stand by passively. It responded economically, attempting to diversify product lines – producing planks, electricity and telephone poles, beams for bridges, railway sleepers, and wooden supports for banana trees. New markets were developed with companies located in areas outside the sphere of influence of local politics.

This process of diversification has continued in recent years. In 1991, the co-operative planned to set up carpentry workshops to train people, in particular adolescents, in more specialized skills. It also planned to set up workshops to train women as seamstresses.

CAFREYOL has taken on the lumber companies by physically reclaiming areas which the companies attempted to take over. During 1989 and 1990, CAFREYOL carried out seven 'recuperations' of areas of forest land 'invaded' by the lumber companies. CAFREYOL also expanded its membership and in 1991 incorporated several peasant groups working exclusively in agriculture. The number of groups affiliated to the co-operative grew to 35 while membership increased from 300 to 1,000.

Despite numerous constraints and pressures, the economic and social situation of the CAFREYOL members has improved since the mid-1980s. The average monthly income of co-operative members in 1990 was approximately $102, three times the minimum wage. There were, however, considerable seasonal variations in wage levels due to increased production levels in summer months when access to forest areas was easier and agricultural activities imposed fewer constraints on members' time. Summer and winter monthly income levels varied from approximately $136 to $68, respectively.

Co-operative members interviewed in the course of this study reported improvements in diet, health, housing and access to basic products through co-operative shops. Some noted that their children now regularly attended classes whereas before they would spend considerable time working the family plot. A literacy programme organized by the co-operative enabled about one thousand adults to learn to read and write. A number of health promoters had been trained in the use of natural medicines while access to medicines in general had improved. The employment opportunities offered by the co-operative had also led to a reduction in the rate of out-migration to work in other areas such as the Aguán Valley where the Standard Fruit company has large plantation interests. Unlike the members of the resin co-operatives which continue to operate in Yoro, CAFREYOL members no longer have to migrate in search of work.

The existence of CAFREYOL had not only brought economic and social benefits for its members but certain environmental benefits for the area in question. Co-operative members had become increasingly aware of the need for forest protection and had undertaken measures to control tree disease, improve the structure of young forests, reduce or eliminate forest fires, and protect river banks and upland areas of watersheds. For these activities the co-operative received some support from COHDEFOR both in terms of technical assistance and food rations obtained from the World Food Programme.

It is apparent that the attitude of co-operative members towards the exploitation of forests had changed significantly during the past five years. There was still, however, a long way to go. Little had been done, for example, in the area of reforestation although in 1991, CAFREYOL did intend to establish a tree nursery and implement a reforestation scheme. In such a scheme there could be an important role for women but there was considerable resistance by men to the involvement of women in productive activities outside of the home. Slash and burn agriculture was still widely practised by co-operative members. While there was more awareness of the negative environmental effects of such practices, co-operative members were unclear as to what alternatives existed.

The existence of the agroforestry co-operative, then, has resulted in significant economic and social benefits for the members who live in an area where alternative employment opportunities are extremely limited. From the standpoint of sustainable natural resource management, there have also been some gains, particularly when compared to the operations of the lumber companies. Far more wood per tree was being used and processed, while a number of sustainable forest management practices were being adopted by co-operative members.

FETRIXY and the Xicaque Indians[4]

The Xicaque Indians constitute another social group which has redefined its relations with the lumber companies, the state and the forests. These changes are closely tied up with transformations that have occurred in the tribal social structure and the institutions of power.

We saw in Chapter 10 that throughout the past century *ladinos* were incorporated into tribal structures as a sort of collective survival strategy. This process of incorporation through marriage with Indian women resulted in a situation whereby those of mixed race, the so-called *indio-ladinos*, came to constitute the majority in many tribes. Yet, historically, this group had been alienated or marginalized within the tribal structure, dominated by the *caciques* who had appropriated relatively large areas of tribal lands for their own benefit, as well as the bulk of the revenue associated with the sale of wood within the SFS. Moreover, the *indio-ladino* group had been marginalized by certain development agencies.

Accompanying these developments was another potentially explosive situation. Over many years, there had been a significant increase in the number of *ladino* families living in tribal areas. As *ladino* communities were formed they came to demand rights to the land. Some also sought municipal status which implied a very different authority structure to that which characterized tribal affairs. Conflicts over encroachment on Indian lands also arose with *ladino* groups in certain areas adjacent to tribal lands where population pressure and inequitable land tenure systems had resulted in a situation of land scarcity.

Over the years, tensions had increased with the group of 'resident *ladinos*'. The Evangelist church, operating through a network of 'preachers' (*celebradores de la palabra*) and agencies such as CARITAS, were active in the Indian communities and gained considerable influence among an emerging group of better educated, younger leaders. Drawn primarily from the ranks of the *indio-ladinos*, this group had considerable contact with some of the more conscientious promoters associated with COHDEFOR and the development project PROCOINY, professionals working on different projects in the department, and certain peasant organizations which were active in the area.

In 1985, this group formed a new organization known as FETRIXY (Federation of the Xicaque Tribes of Yoro) which, in a context where the institution of tribal leadership (*cacicazgo*) had been seriously weakened through corruption, quickly gained considerable influence within many tribes. The Federation's main demands centred on the respect of tribal property rights and the use of natural resources for the benefit and development of the entire community. Many *caciques* were quickly displaced from positions of power and while the *cacicazgo* remained an important position within the tribal power structure, a tribal council was given greater authority. Through elections, FETRIXY supporters or activists came to hold the position of *cacique* in many tribes.

The tribal councils administered tribal property, established criteria for the payment of usufruct rights, and imposed controls on processes of land concentration by establishing ceilings on the area of land which could be used by individual families. They established norms and regulations governing the use and commercial exploitation of forest resources as well as relations with the lumber companies and COHDEFOR. FETRIXY also imposed a tax of between 3 per cent and 6 per cent (often paid in kind) on agricultural production and of 10 per cent of the sale of timber.

The Federation set about recovering Indian lands which had been taken by outsiders and also demanded payment of the stumpage debt which had accumulated over the years with COHDEFOR. To apply pressure on the forestry corporation the tribal councils prohibited the sale of wood.

The wrangling over the debt problem intensified, with COHDEFOR alleging that much of this money had already been paid to the previous *caciques*. In retaliation, COHDEFOR demanded payment of the debt supposedly owned by the tribes for the projects initially implemented as part of the SFS, which FETRIXY claimed had already been paid to the corporation's technical staff.

From the point of view of FETRIXY and the tribes, the conflict with COHDEFOR had both positive and negative implications. While the situation regarding debt repayment remained unresolved, COHDEFOR eventually agreed to increase the stumpage fee from 12 to 20 lempiras although in real terms the increase was probably insignificant. In dollar terms the payment per cubic metre actually fell from $6 in the early 1980s to $4 in 1990.

The conflict put a brake on the process of deforestation and, during the latter half of the 1980s, there was an increase in natural regeneration of pine forest areas. This also resulted from a marked decrease in the numbers of forest fires, which, previously, had often been started as a way of getting revenge on COHDEFOR. FETRIXY also initiated schemes intended to ensure the more sustainable exploitation of forests as well as the manual sawing of wood.

A pilot project in one tribe, for example, aimed to process and market 10,000 cubic metres and eventually set up a small plant to manufacture wooden products.

Despite these developments, the future of the Indian tribes of Yoro remains uncertain. The process of socio-political organization which saw the emergence of FETRIXY has clearly provided many tribes with more power to protect their own resources and exploit them for their own advantage. It has been difficult, however, to overcome the internal divisions and in-fighting of the past. One effect of the boycott was that several FETRIXY activists and the new Indian leaders began to make deals on the side with lumber companies. Often they were unable to fulfil the agreements and several people were subsequently harrassed and imprisoned.

Many of the more 'pure blooded' Indians and the old *caciques* resented the new authority of the FETRIXY activists whom some regarded as 'bureaucrats'. Opposition has also arisen concerning the organization's regulations governing levies. Moreover, several tribes, often those in more isolated areas, were largely by-passed in the FETRIXY programmes.

FETRIXY, however, not only faces opposition from within the Indian sector and from traditional rivals such as COHDEFOR and the lumber companies, but also from a new sector – *ladino* peasant groups. Historically, tensions with this sector were minimized through the process, described earlier, of incorporation via marriage. The formation of FETRIXY, however, put the process of 'ladinization' into reverse gear. 'Incorporation' came to be frowned upon in many tribes. Whereas, historically, it had been vital for the survival of the tribes, a point had been reached where it was threatening tribal rights to land and forest resources and contributing to the disintegration of Indian culture.

DEFORESTATION AND FOREST PROTECTION IN HUEHUETENANGO, GUATEMALA

A short study conducted in the north-western highlands of Guatemala[5] sought to analyse the contrasting situation and responses of two communities in the department of Huehuetenango which were threatened by deforestation. Forests in both areas had been affected by pine borer disease in the 1970s. Certain interest groups had attempted to use this situation as an excuse to clear large areas of forest. In one area, the forces favouring widespread deforestation prevailed; in another, local community residents were able to defend their forests.

In the municipality of San Juan Ixcoy, local businessmen from a nearby town, with the support of the state forestry corporation and the FAO, organized a co-operative to clear 4,000 ha of forest. While much of the local population opposed these activities, which benefited primarily people from outside the community, nothing was done to halt deforestation.

A very different response emerged on the part of the Chuj Indians in San Mateo Ixtatan when they learnt that a lumber company had signed an agreement with the municipal authorities to clear large areas of communal forest. In accordance with tradition, the local authorities were summoned to a public meeting in the town square to explain the nature of the agreement. In response to the opposition of the local population and accusations of corruption, certain authorities, including the mayor, had to resign.

When the company persisted with its plans to clear the forest, protests were organized. A committee of local residents requested the assistance of several national organizations including the Faculty of Agricultural Sciences of the University of San Carlos (which conducted a study showing that only 25 ha had been seriously affected by pine borer disease) and a prominent conservation group. Certain members of the board of the national forestry corporation also supported the residents of San Mateo. As a result of these pressures, the lumber company was obliged to abandon, at least for the time being, its plans to exploit the area.

To explain the very different responses on the part of these two communities, it is important to refer to variations in forms of social organization, culture, economic relations with the forest, and the politico-military situation which prevailed in each area.

The fact that the people of San Mateo mobilized so effectively had to do with several factors. First, it would appear that the local population was more dependent on the use of forest resources, not only for subsistence or basic needs provisioning (fuelwood for cooking, house building materials, etc.) but also as a crucial energy input for salt production which is an important economic activity in the area. Second, traditional forms of communal organization and cultural perceptions regarding the sanctity of the forest and the need to live in harmony with nature were stronger. This was partly due to the historic isolation of the area and relatively limited acculturation. Third, in tactical terms the Chuj Indians adopted an effective strategy. On learning of the threat to their forest, they reacted decisively and not only protested locally but formed an alliance with national institutions which supported their cause.

A very different socio-cultural situation and correlation of social forces existed in San Juan Ixcoy. Here communal structures and traditional cultural beliefs had been gradually eroded through time as a result of a long history of repression and 'ladinization'. While much of the local Indian population opposed deforestation, opposition remained fairly weak, partly due to the weakness of leadership structures. Local leaders remained silent, fearing that any form of protest would be viewed as subversive and repressed by the state security forces. Other community leaders had already left the area due to repression.

As in San Mateo, the question of alliances also proved important in San Juan Ixcoy. Here, however, the dominant alliance involved local entrepreneurs, government forestry officials and certain international development agencies, and favoured deforestation rather than forest protection.

Chapter 12

Social and political dimensions of forest protection

In Central America, as in many Third World countries, the past decade has seen some significant changes in environmental policy and a marked increase in the number of conservation programmes and projects. These developments have been prompted by an upsurge in environmental awareness and in the influence of environmental NGOs, coupled with an increase in environment-linked aid. There are also indications that certain sectors of private enterprise are 'accommodating' to environmentalism and investing in tree plantations, eco-tourism, trade in organically grown products, and so forth. These trends look set to gather momentum throughout the 1990s.

The type of mainstream environmentalism being espoused by governments and leading development agencies, however, has been somewhat limited in scope and beneficiaries. What may be labelled 'environmentalism for nature' and 'environmentalism for profits', tend to hold sway, while a third facet which is essential for sustainable development, namely, 'environmentalism for people' often comes a poor third. This is apparent in various respects which will be elaborated upon below: the failure to integrate many conservation initiatives with the livelihood concerns and priorities of local people; the failure to locate conservation policy and strategy within a coherent development policy framework; and the failure of many agencies to support the organized efforts of local people to defend their natural resource base.

SOCIAL AND POLITICAL ECONOMY CONCERNS

Considerable research has been conducted into the limitations of forest protection and tree planting schemes in Third World countries. There is a tendency, however, to highlight 'technical' aspects associated with the economic and ecological appropriateness of different species, the limitations of extension methodologies, budgetary and bureaucratic constraints affecting the operations of state agencies, price disincentives and so forth. Considerations of social aspects are often reduced to the question of how to motivate farmers and community residents to participate in environmental protection schemes and encourage dialogue between local natural resource users and 'experts'.

One of the central purposes of the Central American study in general, and three of the case studies in particular, was to examine the types of social and political economy constraints that affect project and programme design and implementation. The Costa Rican protected areas study looked at the tensions and conflicts which often emerge when national parks and reserves are created. Another study in Honduras examined the socio-political backdrop which helps to explain the demise of the social forestry system. A third case study in Guatemala, looked at the difficulties confronting environmental projects when they come face to face with the complex social reality which characterizes certain indigenous communities.

Two broad sets of findings emerge from the examination of forest protection and tree planting initiatives in the region. The first relates to questions of policy and project interventions by external agencies; the second, to those of grassroots initiatives and local level institutions and structures.

EXTERNAL INTERVENTIONS

It is apparent that the planning processes associated with the design of conservation policies and programmes are often characterized by a narrow sectoral focus. Insufficient attention is often paid to the broader development context and the linkages between environmental degradation, poverty and capital accumulation.

Governments and certain international agencies tend to follow contradictory development strategies – attempting to encourage environmental protection while at the same time promoting policies and programmes which imply 'business as usual', that is, encouraging patterns of agricultural expansion, industrial and infrastructural development, or mining/logging activities typically associated with unsustainable development. This situation is, to some extent, unavoidable as certain trade-offs will inevitably exist between the biological, economic and social goals generally associated with sustainable development (Holmberg, 1992:24). But there is considerable scope for minimizing trade-offs and, as others have pointed out, to move away from a planning system which attempts to advance in one specific field while disregarding negative impacts in others (ibid.).

The inconsistencies and contradictions which typify development strategies are also intimately tied up with the politics of the policy process and the nature of the state. Far from being a homogeneous entity, the state is highly heterogeneous, with different programmes, divisions or ministries pursuing different goals, promoting different interests, and responding to different constituencies, pressure groups and ideologies.

The danger with this 'multi-track' approach is twofold. First, it results in a 'one step forward two steps back' scenario where, until it is too late, the pace of deforestation is destined to outstrip that of forest protection. Second, it tends to generate certain types of social responses and conflicts which serve to undermine conservation efforts. The failure to address the broader development context is apparent at local, national and international levels.

At the local level, it is necessary to promote appropriate alternatives for local populations whose livelihoods are threatened by environmental rehabilitation initiatives. This is particularly relevant in the case of protected area schemes which displace people or impose restrictions on their traditional subsistence provisioning and economic activities. Moreover, there has been a tendency to focus attention exclusively on protected areas and to disregard the social, economic, cultural and land tenure situation of people in adjacent areas. Insufficient attention has been paid to the development of buffer zones and the question of how to integrate parks and local inhabitants.

For conservation initiatives to minimize the trade-off between environmental protection and human welfare they must involve intensive dialogue with various local groups in the design stage of forest protection or tree planting programmes and projects. Conservation schemes must also attempt to build upon, rather than ignore, the existing stock of local knowledge regarding natural resources and resource management systems. In many areas such systems have been finely tuned over decades or centuries to respond to changing environmental, economic and social circumstances. As a result, they often constitute fairly appropriate methods of 'damage limitation' which serve to minimize both environmental and social risks.

This process of adaptation calls into question the validity of the term 'indigenous or traditional knowledge', which implies something fairly static. As suggested in an UNRISD study of local farmer management systems in Ghana a more appropriate term which captures the element of experimentation with different natural resource management systems might be 'local environmental research' (Amanor, 1992). Whatever term is used, the policy implication is that environmental research must take place on a two-way street with the research establishment learning from, and not simply teaching, farmers and other natural resource users (UNRISD, 1992a:11).

Just as Environmental Impact Assessments (EIAs) are rapidly catching on for many types of development projects, Social Impact Assessments (SIAs) should also become an integral part of the environmental planning process. These would need to determine which social groups stand to win or lose from a specific programme or project, identify potential conflicts of interest, examine procedures for resolving such conflicts, consider how local populations might respond when their livelihood or interests are affected, consider also the environmental and social implications of such responses, and explore alternative livelihood scenarios for individuals and groups affected by such schemes.

This broader perspective linking environmental and human welfare is important, not simply for the obvious reason that it addresses the rights and needs of more marginal groups in society, but also to ensure that fundamental conservation objectives can be achieved in practice. In the Costa Rican study of protected area schemes, for example, it was found that where social conflict had arisen, the state often chose not to implement the law than to add fuel to the fire. Hence non-implementation of protected area status often constituted the means by which to resolve conflicts.

What emerges from the review of the experience of project implementation in Totonicapan, Guatemala is that most of the projects had to learn from bitter experience that technocratic blueprints about what should be done often founder when they come face to face with the real world of complex human relations. Many errors were committed and methodologies had to be modified *en route.*

It is important, however, that such revisions and modifications do take place and that flexibility is built into the project cycle. This is likely to mean that many projects will require longer time frames than is generally the case. As noted in a recent report which summarizes the findings of several UNRISD research projects on socio-environmental issues:

> Participatory and socially accountable implementation procedures are typically time-consuming because of the longer periods necessary for design, impact evaluations and reformulation. They may thus appear less efficient in the short term. A greater degree of flexibility in defining project goals and methods is also necessary to ensure that the project is able to adapt in the best possible way to social and environmental conditions, some of which, in spite of project preparation, may only become evident when the project is in the implementation stage. Most importantly, however, a socially oriented approach will require that projects be less reliant on *ad hoc* interventions and use a more process-oriented approach. It is in this way that projects can move beyond spatially and temporally limited impacts, to be more sustainable in both environmental and social terms. (UNRISD, 1992b:18)

Throughout Central America, project implementation has often been seriously affected by the fact that socio-economic, socio-cultural and socio-political aspects of community life in rural areas have not been adequately understood and incorporated into project design. Reforestation and the promotion of sustainable forest management practices or agroforestry schemes are not simply a question of planting trees or regulating tree felling. Rather they involve changes in the way people use the land and organize their time. They may also involve changes in household income and expenditure patterns as well as the gender division of labour.

Forest protection and tree planting projects can, therefore, affect some fairly fundamental aspects of livelihood and lifestyles. It is crucial that project personnel take on board these types of concerns before diving in at the deep end. Project teams usually lack social scientists who could contribute to the multidisciplinary base of information required for designing appropriate methodologies. Moreover, there is little transfer of the knowledge which is gained from experience. Such experiences are rarely analysed and written up. Many projects are formulated in a vacuum, unaware of the trials and tribulations experienced by other projects.

Budget constraints and the reluctance of many project personnel to be posted in isolated regions, often result in government agencies and NGOs working in areas which are easily accessible. Some rural areas have become saturated with projects. Communities may start to demand of a new project the same incentives offered by another. Moreover, if one or more projects fail to achieve what is promised, community residents can become highly sceptical of projects in general.

In several countries, notably Guatemala, the problem of scepticism felt by much of the indigenous population has to be located in the broader context of Indian–*ladino*, Indian–state relations. The mass of the Indian population in Central America has experienced a history of repression, exploitation and marginalization that has created a culture of mistrust which affects relations with the institutions involved in regulating the use of forest land and forest protection schemes.

The discussion of protected area schemes in Chapter 7 made the point that such attitudes are likely to predominate precisely in the type of outlying forest areas where many protected area schemes are implemented – areas where a history and culture of isolation, 'avoidance', illegality and violence are not uncommon. Such situations require serious consideration by project planners for they raise important questions about the feasibility of protected area schemes, the possibilities and mechanics of dialogue, types of regulations or incentives, levels of compensation, and so forth. Referring to a very different but not unrelated context in West Africa, Richards makes the point that a participatory approach in such situations must place considerable emphasis on working with youth and providing them with an 'alternative vision for the future' (Richards, 1993).

The failure to articulate environmental policies and programmes with broader development questions is particularly evident at the national level where state policies and plans are formulated. Forest protection strategies must go hand in hand with development strategies which attempt to transform specific patterns of accumulation and human settlement which underpin deforestation – notably certain forms of agricultural expansion, infrastructural development, land concentration and the colonization of agrarian frontier areas.

A well-voiced criticism of the numerous Tropical Forestry Action Plans (TFAP) which have been, or are currently being, drawn up in many Third World countries concerns their failure to do just this (Colchester and Lohmann, 1990; Report of the Independent Review, 1990; Winterbottom, 1990). It is notable that only one of the Central American TFAPs – that of Nicaragua – has incorporated, in any systematic manner, this type of analysis.

This approach implies that many types of forest protection and tree planting schemes must not be the sole responsibility of an institute or ministry of natural resources or a conservation NGO. It must, instead, be a joint effort undertaken on a co-ordinated basis by a variety of government, private organizations and foreign aid agencies concerned with economic and social development.

Similarly, the procedures for elaborating forest protection plans should avoid the top-down approach which has tended to characterize this exercise in most countries. Again the Nicaraguan TFAP experience stands out in that it involved systematic consultations with municipal authorities and representatives of NGOs throughout the country.

The case study of Totonicapan, reviewed in Chapter 5, highlighted the linkages between processes of forest destruction on the one hand, and the crisis of subsistence provisioning and the lack of alternative employment opportunities on the other. Until these questions are addressed both through employment-generation and basic needs programmes, the implementation of forest protection schemes is likely to remain, at best, a palliative.

Linking conservation with the broader development context, also means tackling problems associated with land tenure and land rights – problems which prevent a large proportion of Central America's rural population from participating in environmental protection schemes. This is particularly apparent in three respects. First, there is the problem of the small size of holdings. Peasant producers, concerned primarily with investing their time, energy and limited finance in the production of crops and animals which provide the food and income necessary for subsistence provisioning may well prove reluctant to allocate land for tree planting. Second, the failure to respect the customary land rights of indigenous communities can act as a major constraint on conservation initiatives. As witnessed in the case study of Totonicapan, the abuse of communal forest resources by outsiders acted as a disincentive to local residents participating in forest protection schemes when they saw that the fruits of their labour accrued to others. Third, the use and possession of untitled land has also been a factor underpinning deforestation and the failure of certain tree planting schemes. Those with access to untitled land are unlikely to undertake long-term investments in tree planting. Moreover, we have seen how, in certain countries, incentives for reforestation or tree planting schemes are only available for land users with property titles.

Security of tenure, however, should not, as is often the case, be seen as a magic formula which will necessarily produce positive environmental results. As others have pointed out, the line of causality which links security of tenure to environmental rehabilitation via increased investment and productivity can be highly flawed (Bloch, 1992). After all, we should recall that those who acquired security of tenure in the agrarian frontier areas of Central America were often cattle ranchers who in the space of two or three decades destroyed much of the region's forests. Moreover, there is the danger that security of tenure will be associated with the privatization of land and the breakdown of common property regimes. The question of women's rights is also pertinent in situations where titling is likely to benefit men. Who obtains title, how titling affects the rights of different social groups and their access to resources, and the types of farming systems involved, are all questions that need to be examined closely.

In order to implement a conservation strategy which balances concerns for environmental protection with those of human welfare, it may be necessary to affect more powerful vested interests. In the case of the Corcovado National Park in Costa Rica, for example, where conflicts arose between miners panning the rivers for gold and the park authorities, it was necessary to consider affecting the interests of mining companies in order to accommodate those of both the miners and the conservationists. In the case of a country like Guatemala, such an accommodation of interests will probably only be feasible if agrarian reform and land redistribution take place. This is particularly relevant in situations where conservation policy and protected area status are undermined by the constant arrival of land-seeking peasants from other parts of the country.

But while there is a pressing need in much of Central America for major structural reforms and redistributive policies there are no easy answers as is sometimes implied in the so-called 'radical' literature. This is particularly evident with regard to the question of agrarian reform. The experience of both El Salvador and Nicaragua in the 1980s has shown the serious environmental problems which can arise when governments attempt to introduce fundamental structural transformations aimed at improving the situation of the rural poor. The upshot has often been rapid and uncontrolled deforestation in the areas involved. Historically, agrarian reform initiatives throughout Latin America have tended to marginalize or ignore important environmental considerations. Clearly, a new conception of agrarian reform needs to be developed which focuses less on rapid agricultural modernization and land clearance and more on agroforestry or social forestry alternatives, as well as the use of low-cost, low-risk, technologies requiring relatively few imported inputs.

The case study of Río San Juan, Nicaragua revealed the environmental dangers of agrarian reform when the latter focuses exclusively on agricultural expansion and modernization and when credit policy and technical assistance programmes encourage rapid land clearance and input- or capital-intensive techniques. These aspects were partly accounted for by easy access to foreign aid and the pressures of war which demanded the prioritization of short-term productive goals, not least that of food security. They were also characteristic, however, of a certain modernizing approach to development which has been a feature of many societies of socialist orientation (Utting, 1992). The technologies and land-use practices involved proved not only unsustainable from an environmental point of view but also in economic terms once economic stabilization policies set in.

A third aspect concerning the question of policy coherency has to do with the question of North–South relations. Sustainable development processes require not only substantial inputs of international aid but aid consistent with local realities. A coherent international policy environment is also required. In this respect, several contradictions are apparent in the Central American context.

First, national states are being urged by the international community to put their ecological house in order. Yet they have come under tremendous pressure to slash government spending and credit. The fact that the conservationist dynamic of recent years has achieved relatively little on the ground has been due partly to the impact of economic stabilization policies which have led to substantial cuts in certain government expenditures. Such cuts have seriously affected the operations of environmental and extension agencies in most countries. Many state agencies simply do not have the budgets or personnel necessary to fulfil basic counterpart obligations. Moreover, the decline in the real wages of public employees which has occurred in several countries often serves to fuel corruption and 'moon-lighting' and thereby undermine the capacity of the bureaucracy to implement programmes.

Second, structural adjustment programmes promoting the rapid growth of agro-export agriculture have also resulted in contradictory effects. The expansion of non-traditional agriculture in several Central American countries and the recent banana boom, notably in Costa Rica, have often resulted in negative environmental consequences, not only leading to increased deforestation and use of agrochemicals but also destroying more sustainable farming systems and displacing peasant producers to more ecologically fragile areas. It is difficult, however, to generalize about the effects of adjustment programmes on deforestation. To the extent that they promote more intensive forms of 'non-traditional' agriculture, they provide an alternative to the patterns of 'extensive accumulation' which characterized certain types of agro-export agriculture in previous decades and which were a primary cause of deforestation. However, as applied in practice, adjustment programmes, and the fiscal incentives, credit facilities and extension services associated with them, have largely by-passed poorer farmers. As Tucker points out:

> ... policymakers should recognize that, without policy change, non-traditional agricultural production will tend to exaggerate environmental damage as well as income inequities between the rich and the poor. (Tucker, 1992:128)

Perhaps the most glaring contradiction of the 1980s, however, was that associated with the role of the United States government in promoting both conservation and war. While USAID took a leading role in environmental protection initiatives in the region, the United States government also became involved in the internal affairs of certain countries by actively financing, arming and training parties in armed conflict. In countries like El Salvador, the activities of the military resulted in considerable environmental destruction in forest areas. The war in Nicaragua placed the Sandinista government in a strait-jacket which made any movement along a sustainable development path virtually impossible. Budgetary allocations for non-military purposes were increasingly squeezed, conservation was pushed way down the list of government priorities, and access to large forest areas by government and agency officials became virtually impossible.

SOCIAL FORCES AND LOCAL LEVEL STRUCTURES

A more coherent policy framework will not be achieved simply by improving dialogue and communication among policymakers or their awareness of environmental and social issues. A greater degree of policy coherency will ultimately involve changes in the balance of social forces with the emergence of groups or alliances which can challenge the power and influence of traditional élites, constitute new support groups, and bring pressure to bear on policymakers.

It is in this context that the role of NGOs and so-called environmental social movements is important. As Annis points out:

> Previously disconnected groups are forming more complex alliances and political relationships. New forums are emerging for bargaining and deal making. A wider range of social actors is becoming more engaged in economic policy and natural resource issues. Although most decisions are still made to serve the interests of empowered élites, the more complex, modern political environment is generating checks and balances on their prerogatives. (Annis, 1992b:587)

The research revealed how potentially positive state initiatives which attempt to protect the environment and at the same time benefit low income groups are likely to founder if social and political support for such initiatives are not mobilized. Successful policies, programmes and projects require far more than good ideas, technical know-how, material resources and administrative capacity. Certain social and political preconditions must exist. Social and political forces need to be mobilized if change along a sustainable development path is to be effective.

This was brought out clearly in the case study which examined the demise of the Social Forestry System (SFS) in Honduras. The failure of many of the agroforestry co-operatives which were established under the SFS during the mid-1970s is often explained in terms of low prices for pine resin, the administrative inexperience of the co-operatives and the limited capacity of the Honduran state forestry corporation to provide support services. What the research revealed, however, is that the failure of the SFS must also be explained with reference to two other points: a shift in government priorities which reflected changes occurring in the correlation of social forces at the national and international levels; and the reality of social structure and forms of socio-political organization at the local level where projects and programmes were implemented.

The demise of the SFS was not simply due to technical, financial and administrative constraints but occurred within a specific socio-political context. Sectors representing large-scale private enterprise, notably the logging companies, with the backing of the major international lending agencies and the United States government, came to exercise considerable influence within both military and subsequent civilian governments. In policy terms, this was reflected in increased support for the lumber companies and the marginalization of the agroforestry co-operatives.

The Honduran case study also examined the experience of the SFS on the ground, in terms of its impact on, and relations with, peasant and indigenous groups. Here we saw how progressive policies, designed at the central level, can be easily undermined at the local level if they do not have the support of government authorities and dominant interest groups at the sub-national or local levels, or if potential beneficiaries do not actively mobilize in support of such policies. In tribal areas, the SFS was sometimes opposed by Indian chiefs who felt that the new measures undermined their authority. Many chiefs were subjected to considerable pressures by local interests not to comply with regulations governing the SFS. The activities of peasant groups which formed part of the large agroforestry co-operative association, CAFREYOL, were opposed by an array of regional economic, institutional and political forces which included lumber companies which had claims on the area being exploited by the co-operative, the state forestry corporation and departmental and municipal authorities.

The corollary of this analysis is that positive developments involving gains for local populations or specific groups and a more integrative socio-environmentalist approach, often come about through various forms of grassroots organization, mobilization, pressure and struggle. Such actions rarely seek to halt the exploitation of forests but attempt to ensure that the benefits of forest exploitation accrue primarily to local populations and future generations.

Actions on the part of local populations to defend their interests have not only resulted in material benefits for those involved but, in the medium and long term, can induce important modifications in the type of narrow conservationist approach adopted by many governments and agencies. Policymakers do, of course, learn from experience and/or respond to pressure. In a country like Costa Rica with considerable experience in this field, there are definite signs that government agencies and conservation organizations are beginning to take key social questions more seriously. As indicated above, the participatory experience associated with the Nicaraguan planning process during the years of Sandinista rule in the 1980s has, following the change in government in 1990, remained a hallmark of the procedures for drawing up the TFAP.

GRASSROOTS ORGANIZATION AND MOBILIZATION

Much of the contemporary literature on why development projects succeed or fail refers to the need for 'participation', particularly, the need to integrate local populations early on in the project design phase and for institutionalizing mechanisms for ongoing dialogue with project personnel and policymakers.

'Participation', however, involves far more than the active and willing involvement of local people in activities associated with project identification, design, implementation and review. It is also about 'empowerment' or the organized efforts of marginalized groups to transform patterns of resource allocation and increase their control over material resources and resource management decisions.

'Empowerment' is about self-organization: the creation of representative or participatory institutions which enable people to aggregate and articulate interests, mobilize in defense of such interests, or engage in what certain writers refer to as 'environmental action' (Friedmann and Rangan, 1993). As explained by Rangan:

> [t]he idea of environmental action implies several things simultaneously – that people living within any given physical environment assert the need to enhance its quality, to ensure the availability and continuity of the resources they depend upon for their livelihood: and equally important, to cry out a version of their lives to a larger world that is either indifferent or threatens to undermine their attempts at self-definition. (Rangan, 1993)

Several of the Central American case studies demonstrated the importance of: strong forms of grassroots organization and leadership; protest; and the need to construct alliances with organizations or influential individuals operating at the national level. This is in order to bring a local issue to national attention and exert pressure on policymakers.

What also emerges is that in many contexts, grassroots mobilization can serve to win important battles, but the war is never entirely won. The balance of forces between those favouring deforestation and those favouring forest protection is generally in a state of flux. While community-based forest protection initiatives may be effective at one point in time, the pendulum can easily swing back in favour of those who benefit from deforestation. For this reason it is crucial for local groups not only to remain organized and united but also to strengthen links with the growing number of outside agents and organizations with an interest in forest protection.

Much of the so-called 'neo-populist' literature on participation and empowerment pays insufficient attention to the problem of inegalitarian structures, processes of social differentiation, and the skewed distribution of resources at the local level. The Honduran case study of indigenous groups in the department of Yoro brought out clearly the importance of considering such aspects. The study also identified two potentially positive initiatives in terms of sustainable development which have emerged since the mid-1980s in the department of Yoro. One concerned the formation and expansion of the CAFREYOL co-operative. Through an economic strategy involving product diversification and new marketing relations, and a political strategy involving confrontations with lumber interests and alliances with certain national agencies, CAFREYOL was able to advance both in socio-economic and environmental terms. The other involved the creation of the indigenous association, FETRIXY which was able to transform traditional patterns of resource distribution by reducing the power of the *caciques* and negotiating better terms governing the exploitation of forest resources.

The key to transforming highly inequitable patterns of resource use and allocation, in these particular cases, lay in the formation of new forms of organization. Here we see the importance of changes in traditional power structures involving more representative and accountable institutions.

The Honduran case study also indicates the need for what has been referred to as 'mediating institutions' which stand between individuals or households on the one hand, and powerful institutions such as government bureaucracy and capitalist enterprises on the other (Friedmann and Rangan, 1993; Someshwar, 1993). Such organizations would appear to be particularly important in the context of the types of rural settings, often found in Central America, where highly dispersed settlement patterns exist and where there is little social cohesion outside of the family unit. The cases of the Mayan-Quiché Indians in Totonicapán in Guatemala or of the Kuna Indians in Panama, indicate that concentrated settlement patterns and social cohesion have, historically, been important for effective 'environmental action' (Houseal *et al.*, 1985; Veblen, 1978).

As several writers have pointed out, there are hopeful signs that, as democratization takes a hold in Central America, new institutional arrangements are emerging which enable groups with opposed interests to come together, establish dialogue, negotiate and strike deals (Annis, 1992a, 1992b; Arias and Nations, 1992; Cox, 1992). Greater 'citizen participation' (Cox, 1992) is indeed essential if the trade-off between environmental protection and human welfare is to be minimized.

Like security of tenure, democratization, of itself, may do nothing for sustainable development. A country like Costa Rica, for example, has a fairly long tradition of democratic government but has experienced some of the worst environmental degradation in the region. In certain contexts, the emergence or strengthening of formal institutions typically associated with democracy (for example, multi-party structures) may even undermine local level institutions or common property regimes associated with more sustainable forms of natural resource management. But once certain organizational and ideological elements are in place to promote the socio-environmental cause, democracy may provide the type of milieu they need in which to multiply, gain influence and not be repressed.

Participatory democracy may provide a type of enabling environment which makes progress along the path of sustainable development a real possibility. The importance of this aspect can be illustrated with reference to a number of examples provided in a recent book on conservation and development in Central America which highlights several of the region's 'success stories' (Barzetti and Rovinski, 1992). While examples are provided from each of the countries in the region, three in particular stand out for their achievements in the field of conservation or for balancing concerns for environmental protection with those of human welfare. One is the Kuna project in Panama which I referred to in Chapter 11 (Archibold, 1992). Another is a soil conservation project in north-western Nicaragua (Cardenal, 1992). The third concerns the expansion of protected areas and eco-tourism in Costa Rica (Budowski T., 1992). Examples from the other countries tend to be of somewhat isolated or very small-scale experiences, and/or of projects which are still mainly at the planning stage.

It is not entirely coincidental that the more positive developments stem from Costa Rica, Nicaragua and the Kuna 'nation' of Panama, and that the more limited or dubious cases are found in El Salvador, Guatemala and Honduras. The former set of projects have occurred in a specific structural context favourable to environmental action – a context which is all too often forgotten by those concerned with identifying the elements which account for the success or failure of development or conservation projects. This context has been largely absent from the latter set of countries.

The fact that Costa Rica stands out from the other Central American countries for what it has achieved in the field of environmental legislation and programmes, must be understood with reference to the democratic process which exists in that country and the spaces it provides for environmentalists, conservation NGOs and grassroots activists to influence the policy process. Moreover, the democratic process in Costa Rica reflects, to some extent, the peculiar (in Central American terms) structural characteristics of Costa Rican society which saw economic power less concentrated in the hands of a few *latifundistas* or plantation owners, as occurred historically in most other

Central American countries, and more dispersed among a relatively large sector of commercial farmers (Heckadon, 1992:13).

The achievements noted in one particular region of Nicaragua cannot be divorced from the spaces for participation and co-operative organization which emerged following the Sandinista revolution, particularly in the region where the project took place. Not surprisingly perhaps, this project was being threatened by the policies of the new 'neo-liberal' regime which replaced the Sandinista government in 1990 – policies that promoted the expansion of cotton production and reduced credit and other support services to co-operatives.

Finally, the ability of the Kuna Indians to manage their natural resources must be set in the context of the political rights and autonomy they gained following a period of intense struggle in the 1920s and which have enabled them to govern their territory.

The analysis contained in previous chapters has highlighted the complexity of processes of environmental and social change and the need for a holistic and multidisciplinary perspective when analysing such processes. We have seen that the success or failure of environmental protection initiatives is as much about 'technical' as social and political economy aspects.

This study has identified the types of contradictions that arise when conservation policy, programmes and projects fail to address the social dimension or when a coherent policy framework does not exist. It has suggested the need for a more integrative and socially aware approach to environmental planning and project design. It has also indicated the need for agrarian reforms which concern themselves with the promotion of sustainable agriculture or, to adapt a phrase associated with the debate on economic adjustment, agrarian reform with an 'agroforestry face'.

We have also noted, however, the fragility of certain reform experiences, not only from the environmental point of view, but in terms of long-term sustainability. In this respect, reformist top-down approaches are likely to prove short-lived if social and political forces, potentially supportive of such initiatives, are not mobilized, since powerful interests opposed to reforms are likely, in the long run, to hold sway.

I have stressed the opportunities which exist for positive change arising from grassroots mobilization and the need for agencies to be more supportive of such efforts. It is necessary, however, to avoid romanticizing the possibilities for empowerment and participation, given the constraints arising from inegalitarian local structures, co-optation, and ongoing opposition by powerful vested interests. In such contexts, we have seen the importance of democratic processes involving the formation or strengthening of more representative and accountable institutions which can 'mediate' between individuals and the powers that be, as well as of networks and alliances of groups concerned with human rights, basic needs and environmental protection.

When it comes to protecting the world's tropical forests and promoting tree planting in the less developed countries, three aspects of sustainability seem relevant. First, *sustainable conservation* involving the creation of protected areas whose benefits accrue as much to local people as to others such as eco-tourism companies, middle-class tourists, ecologists and the scientific community. Second, *sustainable farming systems*, involving land-use systems based on tree–crop combinations, and technologies which maintain and enhance soil fertility and minimize risk for peasant producers. A third, and more controversial, aspect of sustainability is that of '*sustainable logging*'. Despite the many dangers associated with the concept, certain advantages may be gained from improved incentives and controls on the operations of logging companies, land-use planning or zoning which identifies different forest types and levels of human intervention, and ensuring that local communities participate in the management of forest resources.

What this study has indicated is that progress in these different areas will depend to a large extent on whether or not basic livelihood issues are addressed. Conservation policies and schemes which fail to balance environmental protection and human welfare are likely to result not only in increased hardship for certain social groups but also in forms of social conflict, clandestine activities, non-cooperation or apathy which undermine the possibility of effectively arresting environmental degradation, let alone rehabilitating the natural resource base.

Annex 1: Case studies, theme papers and researchers

Project co-ordinator: Peter Utting

Case studies:

1. La Dinámica Social de la Deforestación en Totonicapán, Guatemala:
   Principal researcher: *Ileana Valenzuela*
   Research collaborators: *César Castañeda*
   Carlos Mendoza

2. Dinámica Social y Deforestación en San Mateo Ixtatán y San Juan Ixcoy, Huehuetenango, Guatemala.
   Researcher: Cesar Castañeda

3. La Dinámica Social de la Deforestación en Yoro, Honduras.
   Researchers: *Alcides Hernández,*
   Santiago Ruíz
   Manéul Chávez

4. Agricultura Migratoria y Deforestación en Río San Juan, Nicaragua.
   Principal researcher: *Emilio Prado*
   Research collaborator: *Rubén Pasos*

5. Social Conflict and Forest Protection in Costa Rica
   Researcher: *Jens Brüggemann*

Theme papers:

1. La Promoción de los Sistemas Agroforestales en América Central
   Consultant: *Carlos Reiche*

2. Políticas y Programas de Protección y Conservación en América Central: Un Creciente Rompecabezas
   Consultant: Juan Carlos Godoy

Notes

INTRODUCTION

1. Services include, for example, the capacity of forests to store carbon, moderate local climates, contribute to soil-building and fertility, control erosion, and regulate hydrological regimes. Goods include timber, pulp, fuelwood, food, animal fodder, medicine and genetic resources. Forests can also have recreational value (Sargent and Bass, 1992a, 1992b).
2. By forest protection and tree planting initiatives I refer to a wide assortment of policies, programmes and projects, as well as individual or collective actions taken at the grassroots level to conserve or rehabilitate forests and/or incorporate trees in farming and livelihood systems.
3. Sustainable development refers to a specific style of 'development' that seeks to promote patterns of accumulation, production, distribution and consumption which minimize the trade-offs that have characterized development processes in most Third World countries, involving economic growth on the one hand, and environmental degradation and social marginalization on the other. It refers to the need to increase levels of human welfare through time and ensure that the well-being of future generations is not compromised by the environmental consequences of today's processes of economic change (WCED, 1987).
4. The term *ladino*, as used in Central America, can have both racial and cultural connotations, referring to persons of mixed (white and Indian) race, or to Indian populations which have experienced considerable 'western' and urban acculturation.

CHAPTER 1

1. This tripartite geographical classification hides a number of important exceptions. Certain Pacific coastal areas in southern Costa Rica and in Panama, for example, experience high rainfall while specific zones in northern Honduras and eastern Guatemala are relatively dry. I am grateful to Gerardo Budowski for these observations.
2. It is likely that this figure includes extremely sparse woodland or areas covered in brush which questionably should be classified as forest.
3. The proportion of total area affected by erosion in Guatemala increases sharply if one excludes the large northern lowland region of the Petén where less erosion occurs.
4. See, for example, Weinberg, 1991.
5. Interview with Andrés Niewenhysen, soils specialist from the University of Wageningen, the Netherlands, working in the northern Atlantic zone of Costa Rica, January, 1991.

CHAPTER 2

1. Interview with Gerardo Budowski, November, 1990.
2. Interview with Gerardo Budowski, November, 1990.
3. An FAO study in El Salvador estimates that a hectare of coffee plantation yields 6.5 cubic metres of fuelwood per annum. The total coffee area yields 1.3 million cubic metres or approximately a quarter of the country's fuelwood requirements (Mansur, 1990).
4. Interview with FAO consultant Antonio Monzón, October, 1990.
5. Interview with Ricardo Navarro, director of the Salvadorean Centre for Appropriate Technology (CESTA), October, 1990.

CHAPTER 3

1. The term 'rich peasant' refers to a category of peasant producers who produce a surplus over and above that which is needed for the subsistence provisioning of the household. This surplus can be used to increase production and/or consumption levels of the household. As Bernstein points out, in so far as they begin to systematically expand production and accumulate capital they come to form a category of capitalist farmers (Bernstein, 1977:67).
2. Interview with Jaime Incer Barquero, September, 1990.
3. Interview with Jaime Incer Barquero, September, 1990.
4. Interview with FAO consultant Antonio Monzón, October, 1990.

PART II OVERVIEW

1. The tragedy of the commons thesis maintains that natural resources in areas of common property are liable to overexploitation since the individuals who access such areas will inevitably act to maximize personal gain to the ultimate loss of the wider community.

CHAPTER 5

1. This section draws to a considerable extent on the report *La Dinámica Social de la Deforestación en Totonicapán, Guatemala,* prepared by Ileana Valenzuela with the assistance of César Castañeda and Carlos Mendoza, under this author's supervision.
2. The UNRISD study confirmed the key role played by communal structures in contributing to forest protection but found little evidence that the organization of carpenters and artisans was an important factor.
3. Personal communication with César Castañeda.
4. Pine borer affects particularly forests where the coloured pine (*Pinus rudis*) predominates and where the lack of selective extraction or thinning leads to excessively dense forest. *Pinus rudis* is a pioneer species which will dominate and restrict the propagation of other species such as white pine in areas which have been clear-felled and where regeneration takes place.
5. Calculated on the basis of 1 lb per capita per day times six family members.
6. See Annex 1.

PART III OVERVIEW

1. The types of proposals or demands put forward by proponents of this position can vary considerably, ranging from 'reformist' to more 'radical' or 'revolutionary' alternatives calling for profound structural change. Such a distinction characterizes environmental social movements in many countries (see García-Guadilla and Blauert, 1992).
2. Personal interview with Steve Mack of CEDARENA and Juan Carlos Godoy who at the time was at CATIE – November, 1990.

CHAPTER 7

1. This country review draws heavily on a report prepared for UNRISD by Juan Carlos Godoy (see Annex 1).
2. Personal interview with Stanley Heckadon, then director of INRENARE, September, 1990.
3. Interview with the minister of IRENA published in *Barricada* newspaper, 24 August, 1991.
4. Part of the information on protected areas in Costa Rica is contained in the report prepared for UNRISD by Jens Brüggemann (see Annex 1).
5. I am grateful to Jens Brüggemann for comments and information he provided about debt for nature swaps in Costa Rica.

CHAPTER 8

1. The section which examines the theme of protected areas and social conflict in Costa Rica draws heavily on a report prepared for UNRISD by Jens Brüggemann (see Annex I).

CHAPTER 9

1. Interview with Manuel Aragón of the company Agroforest SA.
2. Interview with Hans Akersson of Svedforest/Interforest.
3. Interview with José Roberto Denys, Director of the Natural Resources Centre (CENREN) of the Ministry of Agriculture, El Salvador.
4. Interview with Antonio Monzón, FAO consultant.
5. I am grateful to Ramachandra Guha for his comments regarding this point.
6. This review of the agroforestry experience in the region draws on various secondary sources as well as a report prepared for UNRISD by Carlos Reiche: *La Promoción de los Sistemas Agroforestales en América Central* (see Annex 1).
7. The 14 tree species prioritized by CATIE include E.camaldulensis, Eucalyptus saligna, Casuarina equisetifolia, Gliricidia sepium (Madreado), Acacia mangium, Caesalpinia velutina (Aripín), Cupressus lusitanica (Ciprés), Pinus caribaea, Guazuma ulmifolia (Guácimo), Gmelina arborea (Melina), Tectona grandis (Teca), Mimosa scabrella (Bracatinga), Bombacopsis quinatum (Pochote), Leucaena leucocephala.
8. Interview with Carlos Rivas, senior extension officer with CATIE's Madeleña Project.

9. By 1990 some 374 professional and technical personnel had attended short courses on agroforestry. CATIE was also the first institution in Latin America to establish a masters degree in agroforestry (CATIE, 1990) and by 1990 some 91 graduates had obtained masters degrees specializing in agroforestry. Several other institutions in a number of Central American countries also provide courses on aspects associated with agroforestry. Of particular importance is the National School of Forestry Sciences in Honduras which between 1974 and 1989 produced 426 graduates specializing in social forestry.
10. Concerning biological aspects, agroforestry systems permit a better use of vertical space, protection against high rainfall, a reduction in high temperatures and winds. There is a greater yield of organic material, a more efficient recycling of nutrients and improved soil structure. Shade is provided for plants and animals, weeds are reduced, erosion prevented, and there is a greater biodiversity of plant and animal life. Biological disadvantages include the fact that trees may negatively affect crops and animals by competing for water, light, soil nutrients and land. Trees may impede the mechanization of agriculture and encourage certain plants diseases and pests. Tree felling may damage crops.

    Social and economic advantages include the following. Agroforestry may permit a significant diversification of production, enabling the producer to obtain fuelwood, stakes, poles, fruit, animal feed, flowers, and so forth. Agroforestry systems permit the capitalization of the production unit as the value of the trees planted increases through time. Costs of tree planting can be reduced by combining simultaneously tree and crop planting and maintenance activities. Crop production costs may be reduced through reduced incidence of weeds. Fencing costs and the threat to property can be reduced by using trees for live fences. Agroforestry systems create employment, provide work during off-peak periods and enable the workload to be distributed more evenly throughout the year. Socio-economic disadvantages relate principally to the increased demand for labour and the fact that tree planting may diminish the capacity of small peasant producers to grow subsistence crops or cash crops which provide a much quicker return.
11. Interview with Carlos Reiche, senior economist with CATIE's Madeleña project.
12. I am grateful to Solon Barraclough for his comments on cost-benefit analysis.
13. The promotion of eucalyptus in Third World countries has been criticized, often vehemently, on several counts. When grown in association with crops, certain species can, under certain conditions, deprive them of crucial water resources, nutrients and sunlight. Moreover, eucalyptus leaves contain a substance which can negatively affect grazing conditions and the growth of some crops. In a number of countries the promotion of eucalyptus plantations has had the effect of reducing biodiversity not only due to the above effects on plants and animals but also because forest land is often cleared to make way for what is, in effect, a monocrop. In the process, local people's access to essential forest products has been restricted. Moreover, the profitability of eucalyptus plantations in some countries has led landowners to displace peasant producers from the land.
14. Interview with Leonardo Espinoza, GTZ adviser to a Costa Rican agroforestry project in Puriscal, January 1992.
15. Interview with Carlos Rivas, senior extension officer with CATIE's Madeleña Project.
16. Interview with Juan Carlos Godoy, then at CATIE.

CHAPTER 10

1. This review of project implementation in Totonicapan, Guatemala is based largely on a section of the report *La Dinámica Social de la Deforestación en Totonicapán, Guatemala*, prepared for UNRISD by Ileana Valenzuela with the assistance of Cesar Castañeda and Carlos Mendoza under this author's supervision.
2. Agricultural representatives are agricultural producers employed on a part-time basis by the Agricultural Services Division of the Ministry of Agriculture and who act as liaison officers with governmental institutions.
3. In the course of this study we encountered two explanations as to why INAFOR was disbanded and replaced by DIGEBOS. The most commonly accepted explanation is that a new institution was needed to root out corruption and make a fresh start in areas associated with the management of forest resources and forest protection. Another version put to us by former employees of INAFOR, however, claims that a key motive for transforming the institution had to do with the fact that a union organization within INAFOR began to oppose certain policies which threatened the country's forests, particularly those which granted concessions to foreign companies to exploit forests without obliging them to comply with regulations governing reforestation and forest management. The change of institution, then, according to this view became the excuse to get rid of union activists and transfer functions associated with control and application of the law to other state authorities.
4. This examination of the experience of the Social Forestry System in Honduras is based largely on a report prepared for UNRISD by Alcides Hernández, Santiago Ruíz and Manuel Chávez.
5. Interview with Rafael Alegría of the National Council of Rural Workers (CNTC).
6. The case study of the Xicaque Indians was conducted by Manuel Chávez.

CHAPTER 11

1. The Río San Juan study was conducted by Emilio Prado and Rubén Pasos (see Annex 1).
2. This study was conducted by Alcides Hernández and Santiago Ruíz (see Annex 1).
3. Interview with Carlos Rivas, senior extension officer with CATIE's Madeleña project.
4. The study of the Xicaque Indians was conducted by Manuel Chávez (see Annex 1).
5. This study was conducted by César Castañeda (see Annex 1).

Abbreviations and acronyms

ADI	Asociación de Desarrollo Integral; Integral Development Association (Costa Rica)
AMADHO	Asociación de Madereros de Honduras; Honduran Timber Association
AMI	Area de Manejo Integrado; Integrated Management Area (Honduras)
BANSEFOR	Banco de Semillas Forestales; Forest Seed Bank (Guatemala)
CAFREYOL	Cooperativa Agroforestal Regional de Yoro; Regional Agroforestry Co-operative of Yoro (Honduras)
CATIE	Centro Agronómico Tropical de Investigación y Enseñanza; Tropical Agronomy Teaching and Research Centre
CCAD	Comisión Centroamericana de Ambiente y Desarrollo; Central American Commission on Development and the Environment
CECON	Centro de Estudios Conservacionistas; Centre for Conservation Studies
CEDARENA	Centro de Derecho Ambiental y Recursos Naturales; Centre for Environmental Law and Natural Resources (Costa Rica)
CEPAL (ECLAC)	Comisión Económica para América Latina y Caribe; Economic Commission for Latin America and the Caribbean
CESTA	Centro Salvadoreño de Tecnología Apropiada; Centre for Appropriate Technology of El Salvador
CIERA	Centro de Investigación y Estudios de la Reforma Agraria; Centre for Research and Studies of the Agrarian Reform (Nicaragua)
CIM	Comité Intergubernamental para las Migraciones; Intergovernmental Committee for Migration
CNTC	Central Nacional de Trabajadores del Campo; National Council of Rural Workers (Honduras)
COGAAT	Cooperación Guatemalteco-Alemana Alimentos por Trabajo; Guatemalan-German Co-operation for Food for Work
CONAP	Consejo Nacional de Areas Protegidas de Guatemala; National Council for Protected Areas of Guatemala
COHDEFOR	Corporación Hondureña de Desarrollo Forestal; Forestry Corporation of Honduras
COHEP	Honduran Private Enterprise Association
CONAI	Comisión Nacional de Asuntos Indígenas; National Commission for Indigenous Affairs (Costa Rica)
CSD	Cooperación Suiza al Desarrollo; Swiss Development Cooperation.
DEA	Dirección de Educación Ambiental; Environmental Education Directorate (Costa Rica)

DECAFOR	Desarrollo Campesino Forestal; Peasant Forestry Development Programme (Costa Rica)
DEI	Departamento Ecuménico de Investigaciones; Ecumenical Department of Research
DGF	Dirección General Forestal: National Forest Service (Costa Rica)
DIGEBOS	Dirección General de Bosques y Vida Silvestre; Forestry and Wildlife Division (Guatemala)
EEC	European Economic Community
EIA	Environmental Impact Assessment
FAO	Food and Agricultural Organization of the United Nations
FDN	Fundación Defensores de la Naturaleza; Foundation for the Defence of Nature (Guatemala)
FENAGH	National Federation of Honduran Farmers and Ranchers
FETRIXY	Federación de las Tribus Xicaques de Yoro: Federation of the Xicaque Tribes of Yoro (Honduras)
FLACSO	Facultad Latinoamericana de Ciencias Sociales; Latin American Social Science Faculty
FPN	Fundación de Parques Nacionales; National Parks Foundation (Costa Rica)
FTN	Franja Transversal del Norte; Northern Transversal Strip (Guatemala)
GDP	Gross Domestic Product
GNP	Gross National Product
GTZ	Deutsche Gesellschaft für Technische Zusammenarbeit; German Agency for Technical Co-operation
IAN	Instituto Agrario de Nicaragua; Agrarian Institute of Nicaragua
ICE	Instituto Costaricense de Electricidad; Costa Rican Electricity Institute
IDA	Instituto de Desarrollo Agrario; Agrarian Development Institute (Costa Rica)
IFAD	International Fund for Agricultural Development
IICA	Instituto Interamericano de Cooperación para la Agricultura; Inter-American Institute for Agricultural Co-operation
IIED	International Institute for Environment and Development
INRENARE	Instituto de Recursos Naturales Renovables; Institute of Renewable Natural Resources (Panama)
IRENA	Instituto de Recursos Naturales y del Ambiente; Institute of Natural Resources and the Environment (Nicaragua)
ITCO	Instituto de Tierras y Colonización; Land and Colonization Institute (Costa Rica)
IUCN	The World Conservation Union
MAG	Ministerio de Agricultura y Ganadería; Ministry of Agriculture
MIDA	Ministerio de Desarrollo Agropecuario; Ministry of Agricultural Development (Nicaragua)
MIDINRA	Ministerio de Desarrollo Agropecuario y Reforma Agraria; Ministry of Agricultural Development and Agrarian Reform (Nicaragua)
MIRENEM	Ministerio de Recursos Naturales, Energía y Minas; Ministry of Natural Resources, Energy and Mines (Costa Rica)
Mz	Manzana (0.7 ha)

NGO	Non-governmental organization
Q	Quetzal (currency of Guatemala)
qq	Quintal (100 lb)
ROCAP	Regional Office for Central American Programs
SAREC	Swedish Agency for Research and Cooperation with Developing Countries
SECPLAN	Secretaría de Planificación; Planning Secretariat (Honduras)
Sí-a-Paz	Sistema Internacional de Areas Protegidas para la Paz; International System of Protected Areas for Peace
SPN	Servicio de Parques Nacionales; National Parks Service (Costa Rica)
TFAP	Tropical Forestry Action Plan
UNES	Unidad Ecológica Salvadoreña; Salvadorean Ecological Unity
UNICEF	United Nations Children's Fund
UNDP	United Nations Development Programme
UNRISD	United Nations Research Institute for Social Development
UPA NACIONAL	Unión de Pequeños Agricultores; Union of Small Farmers (Costa Rica)
USAID	United States Agency for International Development
WCED	World Commission on Environment and Development
WRI	World Resources Institute
WWF	World Wide Fund for Nature

Bibliography

Alvarado, L. (1985) 'La Sedimentación del Lago de Alhajuela', in S. Heckadon and J. Espinosa (eds), *op.cit.*

Amanor, K. (1992) *Ecological Knowledge and the Regional Economy: Environmental Management in the Asesewa District of Ghana*, paper presented at UNRISD conference on The Social Dimensions of Environment and Sustainable Development, Valletta, Malta, 22–5 April.

Annis, S. (ed) (1992a) *Poverty, Natural Resources and Public Policy in Central America*, Transaction Publishers, New Brunswick.

— (1992b) 'Evolving Connectedness Among Environmental Groups and Grassroots Organizations in Protected Areas of Central America', *World Development*, vol. 20, no. 4, pp. 587–95.

Archibold, G. (1992) 'Pemasky in Kuna Yala: Protecting Mother Earth ... and Her Children', in V. Barzetti and Y. Rovinski (eds), *op.cit.*

Arcia, G., Merino, L. and Mata, A. (1991) *Modelo Interactivo de Población y Medio Ambiente en Costa Rica 1990: Análisis y Proyecciones para el Valle Central*, Asociación Demográfica Costarricense, San José.

Arias, O. and Nations, J. (1992) 'A Call for Central American Peace Parks', in S. Annis (ed), *op.cit.*

Arreaga Garcia, E.H. (1990) *Inventario Forestal y Plan de Manejo del Bosque del Paraje Pacapox, Aldea Chiyax, Municipio de Totonicapán*, Universidad de San Carlos, Quetzaltenango, Guatemala.

Augelli, J. (1987) 'Costa Rica's Frontier Legacy', *The Geographical Review*, vol. 77, no. 1, pp. 1–16.

Baltissen, G. (1989) *Effects of Forest Clearing and Land Use on Soil Properties of Two Land Use Sequences in Cocorí, Atlantic Zone of Costa Rica*, Field Reports No. 34, Atlantic Zone Programme, CATIE/AUW/MAG, Turrialba, Costa Rica.

Barborak, J.R. and Green, G.C. (1985) *Implementing the World Conservation Strategy: Success Stories from Central America and Colombia*, paper presented at the International Symposium on Sustainable Development of Natural Resources in the Third World, September, Ohio State University, Ohio.

Barraclough, S. (1991) *An End to Hunger? The Social Origins of Food Strategies*, Zed Press, London.

— and Scott, M. (1987) *The Rich Have Already Eaten: Roots of Catastrophe in Central America*, Transnational Institute, Amsterdam.

— and Ghimire, K. (1990) *The Social Dynamics of Deforestation in Developing Countries: Principal Issues and Research Priorities*, UNRISD Discussion Paper no. 16, Geneva.

Barrera Arenales, C.A. (1986) *Diagnóstico General de la Aldea Chuculjuyup del Municipio de Totonicapán*, Universidad de San Carlos, Guatemala City.

Barry, D. and Serra, L. (1989) *Diagnóstico Nacional de Nicaragua sobre Refugiados, Repatriados y Población Desplazada 1988*, Cuadernos de Pensamiento Propio, Serie Documentos no. 5, CRIES, Managua.

Barzetti, V. and Rovinski, Y. (eds) (1992) *Toward a Green Central America: Integrating Conservation and Development*, Kumarian Press, London.

Baumeister, E. (1991) *Elementos para Actualizar la Caracterización de la Agricultura Centroamericana*, mimeo, Managua.
Bernstein, H. (1977) 'Notes on Capital and the Peasantry', *Review of African Political Economy*, no. 10, pp. 50–73.
Blaikie, P. and Brookfield, H. (1987) *Land Degradation and Society*, Methuen, London.
Bloch, P. (1992) *Tenure Issues in Forest Buffer Zones*, paper presented at Seminar on Institutional Issues in Natural Resource Management in Africa, 16–20 November, Holbaek, Denmark.
Bogach, V.S. (1981) *A Fuelwood Policy for Guatemala*, Van Meurs and Associates Ltd, Ottawa.
Bonilla Durán, A. (1988) *Crisis Ecológica en América Central*, Ediciones Guayacán, San José.
Boza, M. (1987) 'El Sistema de Parques Nacionales en Costa Rica: Un Ejemplo de Posibilidades para la Conservación en un País en Desarrollo', *Biocenosis*, vol. 3, nos 3–4, pp. 85–101.
Bozzoli, M. (1986) *El Indígena Costarricense y su Ambiente Natural: Usos y Adaptaciones*, Editorial Porvenir, San José.
Bradley, T. *et al.* (1990a) *Costa Rica Natural Resource Policy Inventory*, USAID/ROCAP, vol. II, San José.
— (1990b) *Guatemala Natural Resource Policy Inventory*, (2 vols), USAID/ROCAP, Guatemala City.
Brockett, C. (1990) *Land, Power and Poverty: Agrarian Transformation and Political Conflict in Central America*, Unwin/Hyman, Boston.
Brooijmans, W. (1988) *La Colonización Espontánea de Cocorí, Zona Atlántica de Costa Rica*, Atlantic Zone Programme, CATIE/AUM/MAG, Field Report no. 21, Turrialba, Costa Rica.
— and van Sluys, F. (1990) 'La Lucha por los Recursos Naturales', in Wielemaker W.G. (ed), *Colonización de las Lomas de Cocorí*, Atlantic Zone Programme, CATIE/AUM/MAG, Turrialba, Costa Rica.
Brüggemann, J. (1992) *Population Dynamics, Environmental Change and Development Processes in Costa Rica*, paper presented at UNRISD/FIS conference on The Social Dimensions of Environment and Sustainable Development, Valletta, Malta, 22–25 April.
— and Salas Mandujano, E. (1992) *Population Dynamics, Environmental Change and Development Processes in Costa Rica*, mimeo, UNRISD, Geneva.
Budowski, G. (1980) 'Aspectos Ecológicos del Bosque Humedo', in IRENA (ed), *Primer Seminario sobre Recursos Naturales y del Medio Ambiente*, IRENA, Managua.
— (1981) *Aplicabilidad de los sistemas agroforestales*, paper presented at workshop on Agroforestry in the Humid African Tropics (Ibadam, Nigeria), CATIE, Turrialba.
— (1982) 'The Socio-Economic Effects of Forest Management on the Lives of People Living in the Area: The Case of Central America and Some Caribbean Countries', in E.G. Hallsworth (ed), *Socio-Economic Effects and Constraints in Tropical Forest Management*, John Wiley and Sons, New York.
— (1990) *Deforestation in Central America: Causes, Effects and Possible Ways to Reverse the Trend*, mimeo.
— (1992) 'A Day for Considering Central America', in V. Barzetti and Y. Rovinski (eds), *op.cit.*
— and Veiman, S. (1989) *Misión de Evaluación de la Cooperación Forestal Sueca Swedforest/Interforest en Nicaragua*, mimeo, San José.
Budowski, T. (1992) 'Ecotourism Costa Rican Style', in V. Barzetti and Y. Rovinski (eds), *op. cit.*

van Buren, A. (1990) *The Fuelwood Market in Nicaragua: The Economics, Sociology and Management of a Natural Energy Resource*, Centre for Latin American Research and Documentation, Amsterdam.

Burger, J. (1987) *Report from the Frontier: The State of the World's Indigenous Peoples*, Zed Books/Cultural Survival Inc, London.

de Camino, R. (1989) *La Influencia de Una Comunidad en la Política Forestal de un País. El Caso de Hojancha de Guancaste en Costa Rica*, draft, CATIE, Turriabla, Costa Rica.

Campanella *et al.* (1982) *Honduras: Country Environmental Profile – A Field Study*, USAID.

Cardenal, L. (1992) 'A Tomorrow of Forests and Bread: The Case of Nicaragua', in V. Barzetti and Y. Rovinski (eds), *op. cit.*

Carriere, J. (1991) 'The Crisis in Costa Rica: An ecological perspective' in D. Goodman and M. Redclift (eds), *op. cit.*

Castañeda, C. (1980) *Gorgojo del Pino en Guatemala*, Editorial Universitaria, Guatemala City.

— (1991) *Interacción Naturaleza y Sociedad Guatemalteca: Introducción a su Conocimiento*, Editorial Universitaria, Guatemala City.

CATIE (1981) *Agroforestería*, Turrialba, Costa Rica.

— (1990) *Una Década de Agroforestería en el CATIE*, Turriabla, Costa Rica.

CEDARENA (1990) *Tortuguero Region Case Studies*, mimeo, San José.

CELADE and CIM (1989) *Proyecto de Solicitud de Asistencia al FNUAP: Sistemas Nacionales de Información sobre Poblaciones Desplazadas en América Central*, mimeo, Managua.

Centro Científico Tropical (1988) *Propuesta para la Creación del Sistema Internacional de Areas Protegidas para la Paz (Sí-a-Paz) en el Area del Rio San Juan*, San José.

Centro de Estudios de la Realidad Guatemalteca (CERG) (1989) *Las Selvas de El Petén y el Pulmón Boscoso Mesoamericano: una aproximación a la crisis ambiental de las selvas del norte de Guatemala*, Temas vol. 3, no. 9, Mexico D.F., December.

Centro de Investigaciones y Estudios de la Reforma Agraria, (CIERA) (1981) *La Mosquitia en la Revolución*, Managua.

— (1984) *Nicaragua:... Y Por Eso Defendemos La Frontera*, CIERA-MIDINRA, Managua.

— (1989) *La Reforma Agraria en Nicaragua 1979–1989, Vol.II: Sistema Alimentaria*, Managua.

Centro Nacional de Acción Pastoral (CENAP), (1990) *La Montaña Sagrada: Una Lucha por la Vida*, San José.

Centro Salvadoreño de Tecnología Apropiada, (CESTA), Information brochure, San Salvador.

CEPAL (1991) *Balance Preliminar de la Economía de América Latina y el Caribe 1991*, United Nations, Santiago.

Chacón, I., García, J. and Guier, E. (1990) *Introducción a la Problemática Ambiental Costarricense: Principios Básicos y Posibles Soluciones*, EUED, San José.

Chapin, M. (1987) *The Indians of Guatemala: Problems and Prospects for Social and Economic Reconstruction*, mimeo.

— (1989) *The Indians of El Salvador*, mimeo.

COHDEFOR (1988) *Mesa Redonda 'Participación Internacional en el Desarrollo Forestal de Honduras'*, Tegucigalpa, Honduras.

Colchester, M. (1990a) *Guatemala: The Clamour for Land and the Fate of the Forests*, mimeo.

— (1990b) 'The International Tropical Timber Organization: Kill or Cure for the Rainforests?', *The Ecologist*, vol. 20, no. 5, pp. 166–173.

— and Lohmann, L. (1990) *The Tropical Forestry Action Plan: What Progress?*, World Rainforest Movement, Penang, Malaysia.
Collins, M. (ed) (1990) *The Last Rainforests: A World Conservation Atlas*, Oxford University Press, New York.
CONAP (1990) *Reserva de la Biósfera Maya*, Guatemala City.
— (1990b) *Recopilación e Interpretación de Leyes sobre Areas Protegidas y Conservación de la Naturaleza de Guatemala*, Guatemala City.
Cooperación Guatemalteco-Alemana Alimentos por Trabajo (COGAAT) (1988) *Breve Diagnóstico y Otras Consideraciones sobre los Bosques de Totonicapan*, mimeo, Totonicapán, Guatemala.
— (1989) *Algunas Limitantes de la Producción Agropecuaria Forestal en el Departamento de Totonicapán*, mimeo, Totonicapán, Guatemala.
Cox, S. (1992) 'Citizen Participation and the Reform of Development Assistance in Central America', in S. Annis (ed), *op. cit.*
Current, D. (1991) *Forestry for Sustainable Development: Experiences from Central America and Panama* (final draft), University of Minnesota, St Paul.
Current, K. (1991) *Legislative and Socioeconomic Factors Influencing Rural Land Invasions in Costa Rica*, Research Paper, University of Minnesota, St Paul.
Dickinson, J.C.(ed) (1982) Honduras: Perfil Ambiental del País, Estudio de Campo, McLean, JRB Associates, Virginia.
Diechtl, Sigrid (1988) *Cae una Estrella: Desarrollo y Destrucción de la Selva Lacandona*, Secretaría de Educación Pública (SEP), México D.F.
DIGEBOS (1990) *Proyectos de Reforestación*, MAGA, Guatemala.
Dirección de Educación Ambiental (DEA) (1990) *Parques Nacionales y Areas Afines de Costa Rica 1990*, DEA/SPN/MIRENEM, San José.
Dittborn, A. (1988) *Algunas Actitudes de los Agricultores Hacia la Reforestación en Areas Seleccionadas en Honduras, El Salvador y Guatemala*, CATIE, Turrialba, Costa Rica.
Dudenhoefer, D. (1990) 'Estalla la Polémica', *Aportes*, no. 67, pp. 19–21.
Dulin, P. (1984) *Situación Leñera en los Países Centroamericanos*, CATIE, Turrialba, Costa Rica.
— (1985) *Análisis de Siete Proyectos de Reforestación en Pequeñas Fincas de Ladera en Honduras*, CATIE, Turrialba, Costa Rica.
ECLAC (1982) *Notas sobre la Evolución del Desarrollo Social del Istmo Centroamericano hasta 1980*, Mexico City.
FAO (1977) *FAO Yearbook: Trade 1976*, Rome.
— (1986) *Informe Preliminar: Examen y Análisis de las Políticas y Estrategias para el Desarrollo Rural en Panama*, Rome.
— (1991) *Second Interim Report on the State of Tropical Forests*, mimeo, Forest Resource Assessment 1990 Project, presented at the 10th World Forestry Congress, Paris, September.
— (1988) *FAO Yearbook: Trade 1984, Rome.*
— (1990) *FAO Yearbook: Forest Products 1988*, Rome.
— (1991) *Second Interim Report on the State of Tropical Forests*, Forest Resource Assessment 1990 Project, mimeo, paper presented at the 10th World Forestry Congress, Paris, September.
Finegan, B. and Saboval, C. (1989) 'El Desarrollo de Sistemas de Producción Sostenible en Bosques Tropicales Húmedos de Bajura: Un Estudio de Caso en Costa Rica', *El Chasqui*, CATIE, no. 17, pp. 3–24.
Foley, G. (1987) 'Exaggerating the Sahelian Woodfuel Problem', *Ambio*, no. 16 (6), pp. 367–71.
Fournier, L.A. (1991) *Desarrollo y Perspectiva del Movimiento Conservacionista Costarricense*, Editorial de la Universidad de Costa Rica, San José.

Friedmann, J. and Rangan, H. (eds) (1993) *In Defense of Livelihood: Comparative Studies in Environmental Action*, Kumarian Press, Westford, C.T.

Fundación Defensores de la Naturaleza y El Fondo Mundial para la Vida Silvestre (FDN/WWF) (1989) *Estudio Técnico para Dar a Sierra de las Minas la Categoría de Reserva de la Biósfera*, Guatemala City.

Gallardo, M.E. and López, J.R. *Centroamérica: La Crisis en Cifras*, IICA/FLACSO, San José.

García-Guadilla, M-P. and Blauert, J. (eds) (1992) *Environmental Social Movements in Latin America and Europe: Challenging Development and Democracy*, MCB University Press, Hull.

Gente (1990) (supplement of *Barricada* newspaper), Managua, September.

Gewald, N. (1980) *The Importance of Fuelwood in Central America: an appraisal and plan for action*, mimeo, CATIE, Turrialba, Costa Rica.

Godoy, J.C. (1990) *Riqueza y Pobreza Extremas: el Ambiente Natural y Los Recursos*, mimeo.

— (1992) 'Monterrico: A Reserve for Maximum Use', in V. Barzetti and Y. Rovinski (eds), *op. cit.*

Goodman, D. and Redclift, M. (1991) Editors' Introduction in Goodman and Redclift (eds), *Environment and Development in Latin America: The politics of sustainability*, Manchester University Press, Manchester.

Gradwohl, J. and Greenberg, R. (eds) (1988) *Saving the Tropical Forests*, Earthscan Publications, London.

Gramajo Castillo, E.V. (1980) *Monografía del Paraje de la Zona de Palin del Departamento de Totonicapán*, Universidad de San Carlos, Quetzaltenango, Guatemala.

Granados Loarca, J.N. (1989) *Inventario Forestal y Plan de Manejo del Bosque del Cerro Alaska, Cantón Patzite, Municipio de Nahualá, Departamento de Sololá*, Universidad de San Carlos, Guatemala.

Green, G. (1990) 'Conservation through the looking glass: the case of Central America', in M. Palo and G. Mery (eds), op.cit, pp. 121–9.

Guatemalan government (1978) *Agricultural Basic Grains Survey*, Guatemala City.

Guatemalan government (1990) *La Contribución del Sector Forestal al Desarrollo*, Plan de Acción Forestal para Guatemala, Guatemala City.

Hall, B. and Faber, D. (1989) *El Salvador: Ecology of Conflict*, The Environmental Project on Central America, EPOCA, Green Paper no. 4.

Hartshorn, G. *et al.* (1982) *Costa Rica: Perfil Ambiental*, Tropical Science Center, San José.

Healy, R. (1988) 'Problemas de Desarrollo y de la Conservación de los Recursos Naturales en Centroamérica', in W. Ascher and A. Hubbard (eds), *Recuperación y Desarrollo de Centroamérica*, Duke University, San José.

Heckadon Moreno, S. (1982) 'La Colonización Campesina de Bosques Tropicales en Panama', in S. Heckadon and A. McKay (eds), *op. cit.*

— (1984) *Panama's Expanding Cattle Front: The Santeño Campesinos and the Colonization of the Forests*, Ph.D. thesis, University of Essex, Colchester.

— (1985) 'La Ganadería Extensiva y la Deforestación: los Costos de una Alternativa de Desarrollo', in S. Heckadon and J. Espinosa (eds), *op.cit.*

— (1989) 'Los Viveros Comunales en El Salvador', *El Chasqui* no. 20, pp. 3–24.

— (1990) *Madera y Leña de las Milpas. Los viveros comunales: una alternativa para el desarrollo forestal en El Salvador*, CATIE, Turrialba, Costa Rica.

— (1992) 'Central America: Tropical Land of Mountains and Volcanoes', in V. Barzetti and Y. Rovinski (eds), *op. cit.*

— and Mckay, A. (eds) (1982) *Colonización y Destrucción de Bosques en Panamá*, Asociación Panameña de Antropología, Panama City.

— and Espinosa, J. (eds) (1985) *Agonia de la Naturaleza*, INDIAP/Smithsonian Tropical Research Institute, Panama City.

Hedström, I. (1988) *Somos Parte de un Gran Equilibrio: La Crisis Ecológica en Centroamérica*, DEI, San José.

— (ed) (1989) *La Situación Ambiental en Centroamérica y el Caribe*, DEI, San José.

— (1990) *Volverán las Golondrinas?: La Reintegración de la Creación desde una Perspectiva Latinoamericana*, DEI, San José.

Hernández, A. (1982) 'Migración de Colonos en Darién', in S. Heckadon and A. McKay (eds), *op.cit.*

— (1986) *Subsistema Forestal de Honduras*, UNAH, Tegucigalpa.

— and Ruíz, S. (1991) *Deforestación y Cooperativas Agroforestales: El Caso del Departamento de Yoro, Honduras*, mimeo, Tegucigalpa.

Holmberg, J. (ed) (1992) *Policies for a Small Planet*, Earthscan, London.

Houseal, B., MacFarland, C., Archibold, G. and Chiari, A. (1985) 'Indigenous Cultures and Protected Areas in Central America', *Cultural Survival Quarterly*, March, pp. 10–20.

Howard, P.(1987a) *From Banana Republic to Cattle Republic: Agrarian Roots of the Crisis in Honduras*, Doctoral Thesis, Ann Arbor.

— (1987b) *Frontier Expansion, Deforestation and Agrarian Social Change: The 'Cattleization' of the Nicaraguan Countryside, 1950–1971*, mimeo, Universidad Nacional Autónoma de Honduras, Tegucigalpa.

IIESO (1976) *Guatemala: Estructura Agraria del Altiplano Occidental*, Centro Universitario de Occidente, Guatemala.

Instituto Agrario de Nicaragua (IAN) (1972) *Proyecto Rigoberto Cabezas I Etapa*, Managua.

Instituto Interamericano de Cooperación para la Agricultura (IICA) and Facultad Latinoamericano de Ciencias Sociales (FLACSO) (1991) *Centroamérica en Cifras*, IICA, San José.

Instituto Nacional de Recursos Naturales Renovables (INRENARE) (1991) *Plan de Acción Forestal de Panamá*, Documento Principal, Panama City.

International Fund for Agricultural Development (IFAD) (1984) *Estrategia para el Desarrollo de la Producción Campesina en el Altiplano Occidental*, Informe de Misión Especial de Programación a la República de Guatemala, Rome.

IRENA *et al.* (1985) *Plan de Desarrollo Forestal de la República de Nicaragua*, Informe Principal, Stockholm.

Jenkins, J. (1986) *El Desafío Indígena: El Caso de los Miskitos*, Editorial Catún, Bogotá.

Jones, J. (1982) *Socio Cultural Constraints in Working With Small Farmers in Forestry: Case of Land Tenure in Honduras*, mimeo, CATIE, Turrialba, Costa Rica.

— (1985) *Land Colonization in Central America*, CATIE, Turrialba, Costa Rica.

— (1988) 'Colonization in Central America', in W. Manshard and W. Morgan (eds), *Agricultural Expansion and Pioneer Settlements in the Humid Tropics*, UNU, Tokyo.

Kapp, G. (1989) *Perfil Ambiental de la Zona Baja de Talamanca, Costa Rica*, CATIE/GTZ/DGF, Turrialba, Costa Rica.

Kauck, D. and Tosi, J. (1990) *Public Policy Decisions and Private Resource Use: Changing Patterns of Deforestation and Land Use in the Arenal Basin*, Tropical Science Centre, San José.

Larson, A. (1989) 'Ecología y Política: Los Problemas "Verdes"', *ENVIO*, no. 97, September.

Leach, G. and Mearns, R. (1988) *Bioenergy Issues and Options for Africa*, IIED, London.

Leal Ruíz, M.A. (1989) *Inventario Forestal y Plan de Manejo del Bosque Comunal del Centro de Nahualá y de los Cantones Xepatuj y Chuirjraxom*, Universidad de San Carlos, Quetzaltenango, Guatemala.

Leonard, H.J. (1985) *Recursos Naturales y Desarrollo Económico en América Central: Un Perfil Ambiental Regional*, IIED, San José.

— (1987) *Natural Resources and Economic Development in Central America: A Regional Environmental Profile*, International Institute for Environment and Development, Transaction Books, New Brunswick.

Lohmann, L. and Colchester, M. (1990) 'Paved With Good Intentions: TFAP's Road to Oblivion', *The Ecologist*, vol. 20 no. 3, May/June.

Mahony, R. (1992) 'Debt-for-nature swaps: Who really benefits' *The Ecologist*, vol. 22, no. 3, pp. 97–103.

Major, M., Reiche, C. and McKenzie, T. (1989) 'Participación de la Mujer en la Reforestación de Hojancha, Costa Rica', *Silvoenergía*, CATIE, no. 29, Turrialba, Costa Rica.

Mansur, E. (1990) *El Salvador: Plan Nacional de Reforestación de El Salvador*, FAO, San Salvador.

Marroquín, A. (1975) 'El Problema Indígena en El Salvador', *América Latina*, vol. 35, no. 4, pp. 747–71.

Martínez, H. (1986) *El Problema de la Leña en las Zonas Secas de América Central: Necesidades de Investigación*, mimeo, CATIE, Turrialba, Costa Rica.

—, Bauer, J. and Jones, J. (1983) *Fuelwood in Central America and the Regional Fuelwood and Alternative Energy Sources Project*, CATIE, Turrialba, Costa Rica.

McBryde, F. (1969) *Geografía Cultural e Histórica del Suroeste de Guatemala*, Guatemela.

McCreery, D. (1976) 'Coffee and Class: The Structure of Development in Liberal Guatemala', *Hispanic American Historical Review*, vol. 56, no. 3, pp. 438–60.

McGaughey, S. and Gregersen, H. (eds) (1983) *Forest-Based Development in Latin America*, Inter-American Development Bank, Washington D.C.

McGhie, J. (1987) 'Reclaiming a Natural Legacy', *The Ecologist*, vol. 17, no. 4/5, pp. 200–202.

McKay, A. (1982) 'Colonización de Tierras Nuevas en Panamá', in S. Heckadon Moreno and A. McKay (eds), *op.cit.*

Merino, L. and Mata, A. (1991) *Proyecto Regional de Población y Medio Ambiente: Caso de los Recursos Naturales Renovables*, Plan de Acción Forestal para Centroamérica, San Jose.

Ministerio de Agricultura y Ganadería (MAG) de Nicaragua (1990) *Incremento de la Productividad Agropecuaria y Conservación de Recursos Suelo, Aguas y Bosques*, Managua.

Ministerio de Desarrollo Agropecuario (MIDA) de Nicaragua, (1980) *Proyecto Desarrollo Rural Integral Rigoberto Cabezas II Etapa*, vol. I, Managua.

Ministerio de Recursos Naturales, Energia y Minas (MIRENEM) de Costa Rica (1990) *Plan de Acción Forestal para Costa Rica: Documento Base*, San José.

Mora, J.A. (1991) *Generación de Empleo en Zonas Agroforestales Seleccionadas de Costa Rica*, FAO, Rome.

Morales, R. and Cifuentes, M. (1989) *Sistema Regional de Areas Silvestres Protegidas en América Central: Plan de Acción 1989–2000*, CATIE, Turrialba, Costa Rica.

Myers, N. (1990) 'Guardianes de la Selva Panameña', *Reader's Digest Selecciones*, November, pp. 78-84 (abridged version of article published in *International Wildlife*, July/August, 1987).

Nations, J. and Komer, D. (1987) 'Rainforests and the Hamburger Society', *The Ecologist*, vol. 17, no. 4/5.

Nations, J. and Leonard, H.J. (1986) 'Grounds for Conflict in Central America', in A. Maquire and J. Welshbrown (eds) *Bordering on Trouble*, Alder and Alder, Boston.
Nuñez, R., Serrano, F., Martínez, A.C. and Guerra, H. (1990) El Salvador Natural Resource Policy Inventory (draft), vol. 1, synthesis, USAID/ROCAP RENARM Project.
Palo, M. and Mery, G. (eds) (1990) *Deforestation or Development in the Third World?*, Bulletin 349 of the Finnish Forest Research Institute, vol.III, Helsinki.
Pedroni, L. (1991) 'Conservación y Producción Forestal: Aspectos para su Conciliación en el Marco de un Manejo Sostenible', *El Chasqui*, no. 27, November, pp. 7–22.
Peters, R. (1985) *Estimación de la Deforestación en Nicaragua*, mimeo, IRENA, Managua,
Pickles, D. (1992) 'La Batalla Verde', *Pensamiento Propio*, vol. 10, no. 90, May, pp. 12–13.
Plan de Acción Forestal Tropical para Centroamérica (1991) *Taller Centroamericano sobre Manejo de Bosque Latifoliado de Bajura*, CATIE, Turrialba, Costa Rica.
Poole, P. (1990) *Desarrollo de Trabajo Conjunto entre Pueblos Indígenas, Conservacionistas, y Planificadores del Uso de la Tierra en América Latina*, CATIE, Turrialba, Costa Rica (translation of 1989 World Bank report).
Poore, D. (ed) (1989) *No Timber without Trees: Sustainability in the Tropical Forest*, Earthscan Publications, London.
Rafael Landivar University Institute of Environmental Sciences and Agricultural Technology (ICATA) (1984) *Environmental Profile of Guatemala*, URL/AID-Guatemala/ROCAP, Guatemala City.
Rangan, H. (1993)'Romancing the Environment: Popular Environmental Action in the Garhwal Himalayas', in J. Friedmann and H. Rangan (eds) *op. cit.*
Reiche, C. (1983) *Implicaciones Económicas del Componente Agroforestal*, mimeo, CATIE, Turrialba, Costa Rica.
— (1986) *La Leña en el Contexto Socioeconómico de América Central*, mimeo, CATIE, Turrialba, Costa Rica.
— (1988) *Socio-Economic Approach and Analysis of Agroforestry Systems Applied on Demonstration Farms in Central America*, paper presented at symposium Fragile Lands in Latin America: The Search for Sustainable Uses, XIV Congress of the Latin American Studies Association, 17–19 March, New Orleans.
— and van Buren, A. (1984) *Wood Fuel Commerce in Nicaragua*, mimeo, CATIE, Turrialba, Costa Rica.
Report of the Independent Review (1990) *Tropical Forestry Action Plan*, Kuala Lumpur.
Revista Del Campo (1990) (supplement of *Barricada* newspaper), 12 January, Managua.
Richards, P. (1993) *Indigenous Peoples*, Thematic Issue Paper of the Proceedings of the Royal Society of Edinburgh on the Lowland Rain Forest of the Guinea-Congo Domain, mimeo.
Riding, A. (1979) 'Guatemala Opening New Lands But the Best Goes to Rich', *New York Times*, p. A1.
Rivas, C. (ed) (1988) *Memoria de la Conferencia de Especialistas en Extensión – América Central*, CATIE, Turrialba, Costa Rica.
Rodriguez, S.C. and Vargas, E. (1988) *El Recurso Forestal en Costa Rica: Políticas Públicas y Sociedad, 1970–1984*, Editorial de la Universidad Nacional, Heredia, Costa Rica.
Rubinoff, I. (1982) 'Los Problemas Ambientales que Confronta Panamá' in S. Heckadon and A. McKay (eds), *op. cit.*
Sargent, C. and Bass, S. (eds) (1992a) *Plantation politics: Forest plantations in development*, Earthscan, London.

— (1992b) 'The Future Shape of Forests', in J. Holmberg (ed), *op.cit.*

Sarmiento, M. (1985) 'Meteti: Una Comunidad que Abre la Selva del Darién', in S. Heckadon and J. Espinosa (eds), *op.cit.*

SECPLAN, DESFIL and USAID (1990) *Perfil Ambiental de Honduras 1989*, Tegucigalpa.

SEGEPLAN (1991) *Análisis Interpretativo de la Situación Actual del Departamento de Totonicapan*, mimeo, Guatemala City.

Shaw, P. (1989) 'Rapid Population Growth and Environmental Degradation: Ultimate and Proximate Factor', *Environmental Conservation*, vol. 16, no. 3, pp. 199–208.

Silliman, J. (ed) (1981) *Draft Environmental Profile of the Republic of Costa Rica*, Arid Lands Information Center, Tuscon.

Simons, P. (1988) 'Costa Rica's Forests are Reborn', *New Scientist*, 22 October.

Smith, C. (1990) 'The Militarization of Civil Society in Guatemala: Economic Reorganization as a Continuation of War', *Latin American Perspectives*, Issue 67, vol. 17, no. 4, Fall, pp.8–41.

Solis, P. (1989) 'The Atlantic Coast of Nicaragua: Development and Autonomy', *Journal of Latin American Studies*, vol. 21, pp. 481–520.

Someshwar, S. (1993) 'The Social Forestry Program in Karnataka State, India: People Versus the State?, in J. Friedmann and H. Rangan (eds), *op.cit.*

Synnott, T. (1989) 'South America and the Caribbean', in D. Poore (ed) *op.cit.*

Thrupp, L.A. (1980) *Deforestation, Agricultural Development and Cattle Expansion in Costa Rica*, Honours Thesis, Human Biology Programme, Latin American Studies, Stanford University, California.

Tschinkel, H. (1984) *Tree Planting by Small Farmers in Upland Watersheds: Experience in Central America*, paper presented at IX World Forestry Congress, Mexico.

Tucker, S. (1992) 'Equity and the Environment in the Promotion of Nontraditional Agricultural Exports', in S. Annis (ed), *op. cit.*

UNDP, (1991 and 1992) *Human Development Report*, Oxford University Press, New York.

UNICEF (1988) *Infancia y Guerra en El Salvador*, Guatemala City.

Unidad Ecológica Salvadoreña (UNES) (1990) *Propuesta del Cerro Verde*, San Salvador.

United Nations *Statistical Yearbook*, 1970, 1975, 1983, 1984, United Nations, New York.

UNRISD (1979) *UNRISD: Studies for Social Change*, Geneva.

— (1986) *Food Systems and Society: Problems of Food Security in Selected Developing Countries*, Geneva.

— (1990) *The Social Dynamics of Deforestation in Developing Countries: Issues and Research Priorities*, paper presented to workshop on Social Dynamics of Deforestation in Developing Countries, Geneva, 27–29 August.

— (1992a) *Social Development Research: UNRISD Activities 1991/92*, Geneva.

— (1992b) *Development, Environment and People*, Geneva.

USAID (1985) *El Salvador: Perfil Ambiental de Campo.*

Utting, P. (1992) *Economic Reform and Third-World Socialism*, Macmillan, London.

Vahrson, W-G. and Cervantes, C. (1991) 'Escorrentía superficial y erosión laminar en Puriscal, Costa Rica', in W-G. Vahrson, M. Alfaro and G. Palacios (eds), *Taller de Erosión de Suelos*, Memoria, Universidad Nacional, Heredia, Costa Rica.

Valenzuela, I. (1989) *Technologie Appropriée, Quelle Alternative pour le Développement?* Postgraduate Thesis, Institut Universitaire d'Etudes du Développement, Geneva.

Veblen, T. (1978) 'Forest Preservation in the Western Highlands of Guatemala', *The Geographical Review*, vol. LXVIII, pp. 417–34.

Vilas, C. (1989) *Transición desde el Subdesarrollo: Revolución y Reforma en la Periferia*, Editorial Nueva Sociedad, Caracas.
Vivian, J. (1991) *Greening at the Grassroots: People's Participation in Sustainable Development*, Discussion Paper no. 22, UNRISD, Geneva.
Vogt, W. (1946) *The Population of El Salvador and Its Natural Resources*, Pan American Union, Washington D.C.
Wadsworth, F. (1982) 'La Deforestación, Muerte del Canal de Panamá, in S. Heckadon and A. McKay (eds), *op. cit.*
Weeks, J. (1985) *The Economies of Central America*, Holmes & Meier, New York.
Weinberg, B. (1991) *War on the Land: Ecology and Politics in Central America*, Zed Press, London.
Whelan, T. (1988) 'Central American Environmentalists Call for Action on the Environment', *Ambio*, no. 17, pp. 72–5.
Williams, R. (1986) *Export Agriculture and the Crisis in Central America*, University of North Carolina Press, Chapel Hill.
Winterbottom, R. (1990) *Taking Stock: The Tropical Forestry Action Plan After Five Years*, World Resources Institute, New York.
Woodward, R. (1985) *Central America: A Nation Divided*, Oxford University Press, Oxford.
World Bank (1989) *World Development Report 1989*, Oxford University Press, New York.
— (1992) *World Development Report 1992: Development and the Environment*, Oxford University Press, New York.
World Commission on Environment and Development (WCED) (1987) *Our Common Future*, Oxford University Press, New York.
Wunderlich, V. and Salas Mandujano, E. (1991) *Conflictos de Uso en las Zonas Periféricas: El Impacto de la Colonización en los Territorios Indios en el Sur de Costa Rica*, ASA Programme, Berlin.
Zanotti, J. (1986) *Implicaciones Sociales de los Programas de Plantaciones para Leña en Guatemala*, mimeo, CATIE, Turrialba, Costa Rica.

Index

acculturation 16, 53, 159
ADI (Integral Development Associations) 104
agrarian frontier regions 26, 41, 43, 48, 49; colonization 17–19, 25, 34, 36, 38, 98, 164
agrarian reform 42–3, 133, 140, 147–50; environmental dangers of 166; initiatives 166; need for 172; programme 29, 115, 139
agrarian structure 5; and land tenure 34–7
Agricultural Basic Grains Survey 72
Agricultural Censuses 73
agriculture 11, 59, 74; agro-export 34, 43, 167; areas cleared for 62; basic wage 78; commercial 12, 30, 36, 125; domestic 54; employment opportunities 29; migratory 42, 52, 84–8; non-traditional 167; peasant 12–13, 20, 84, 72–4, 86; rapid modernization 166; rudimentary 144; shifting 56, 82–91, 149; slash and burn 16, 84, 156; soils not suitable for 73; subsistence 26, 84; sustainable 112, 172; traditional practices 53; value added in 44; *see also* production; productivity
agrochemicals 25, 64, 73, 149, 167
agroforestry 25, 49, 86, 166; co-operatives 138, 139, 142–3, 146, 168; land-use systems 146; projects 133–7; reform 43, 149; research and dissemination 89; rigid formulas regarding practices 132; schemes 14, 42, 110, 118–30, 163
agro-industrial development 18
Aguán Valley 155
aid agencies 81, 130; bilateral 94, 113; foreign 125, 134; international 119; *see also* CSD; USAID
alcohol 54, 112
Alvarado, L 10
AMADHO (Honduran Lumber Dealers Association) 140, 141
Amanor, K 162
Amazonian region 13
AMI (Areas of Integrated Management) 138, 139, 154
Annis, S 89, 167–8, 171
apiculture 145
Archibold, G 151, 171
Arcia, G 10, 15, 22, 30
Arenal Forest Reserve 106, 107
Arias, O 89, 171
aridity 11
Arreaga Garcia, E H 60
artisanry 22, 59, 145
ash deposits 82
Atlantic coast regions: Costa Rica 13, 19, 22, 26; Honduras 144; Nicaragua 49, 51, 54, 85–6
Augelli, J 36, 39–40
autonomy 151, 172
Azuero Peninsula 12, 18

bananas 15, 23, 25–6, 55, 140, 167; floods in plantations 12
Banco Nacional de Desarrollo 38
bank collateral 111
Barborak, J R 49
bark stripping activities 64–6, 137
Barra Colorado 93; Wildlife Refuge 105, 108
Barraclough, S 35, 36, 150
barreras 73
Barricada (newspaper) 31, 49
Barry, D 41
Barzetti, V 171
Bass, S 115
Baumeister, E 5, 43, 44

bee-keeping activities 110, 145
beef 15, 19, 28, 34, 43
Belize 48n, 98
bilateral/multilateral development agencies 29; *see also* CSD
biodiversity 11, 74, 100, 109, 115
biomass 82
black pepper (*pimienta gorda*) 98
Blaikie, P 15
Bloch, P 165
Boaco 33
Bogach, V S 30
Bonilla Durán, A 9n
booms 54; banana 55, 167; cotton, beef and sugar 34; lumber 55
Borunca reserve 51
Bosawas forest reserve 94
BOSCOSA project 117
botanical sanctuary 151
Boza, M 101
Bozzoli, M 48n, 49
Bradley, T 6, 9n, 11, 41, 50, 97, 103, 114, 117
Bragman's Bluff Lumber Co 54
Brazil 13
bribes 80, 137, 145
Brockett, C 23, 25, 26, 40, 114
Brooijmans, W 19
Brookfield, H 15
Brüggemann, J 11–12, 13, 19, 22, 26, 83, 101, 105, 106, 126
Budowski, G 21–2, 27, 37, 89, 119, 130
Budowski, T 171
buffer zones 162
Burger, J 48n, 50, 51, 151
burning activities 12, 74; *see also* slash and burn

Cabañas 41
caciques (tribal leaders) 144, 145, 156, 157, 158, 170
CAFREYOL (Regional Agroforestry Co-operative of Yoro) 142, 143, 153–6, 168, 170
capital 5, 17, 26, 44, 161
Carara Reserve 109–11
Cardenal, L 127, 171
CARE (Co-operative for American Relief) 119, 134, 135
cargas 78
Caribbean coastal/lowland areas 1, 11, 25, 151
carpentry industry 72, 74–7, 81
Carriere, J 34
cash crops 23, 51, 64, 73, 110
Castañeda, C 63
Catholic church 139, 140
CATIE studies 30, 118–120, 124, 127, 129, 132, 151, 154
cattle 12, 19–21, 38, 83, 129; expansion 114, 142; grazing land for 10, 53, 54, 87, 138, 150; large extensive farms 5; major producer for export 39; raising 14, 84, 86, 87; ranching 19, 98–9; soil compactation due to the movement of 11
cedar 98, 132
CEDARENA report (1990) 36, 40, 103
Central American Commission on Environment and Development 89
Central American Common Market initiative 27
CEPAL report (1991) 5, 102
Cervantes, C 13
Chacón, L 10
Chalatenango 41
Chapin, M 18, 48n, 50
chicle (gum) 98
child care 33
children 33, 55, 60, 78, 121, 122
Chimaltenango 77
Chinandega 31
Chiriqui 18
Chirripo National Park 108
Choloteca Valley 13
Christmas trees (*Pinus abeis*) 70
Chuj Indians 158
CIERA (Centro de Investigaciones y Estudios de la Reforma Agraria) 54, 82, 84, 85–6, 87
Cifuentes, M 92
clandestine economy 68, 81, 98, 173
climate 1, 86; micro- 10, 31, 74; temperate 1, 130
coastal fishing communities 32
coercion 36, 53
coffee 5, 15, 23–5, 51, 84, 85, 126; encroachment of plantations 151; expansion 16; falling prices on world market 25; growth of production 34; harvests 72; producing regions 132;

suitable tree densities in plantations 119; zones 33
COHDEFOR (Honduran Forestry Development Corporation) 45, 137–41, 143–5, 152–5 *passim*, 157–8
Colchester, M 117, 164
Collins, M 93
Colombia 18, 50, 93
colonization *see* agrarian frontier colonisation; land colonization
coloured pine (*Pinus rudis*) 66
commerce 59
commercial gain 64, 67
communal holdings 34
'communist' measures 140
community-based protection mechanisms 66–7, 80
compensation 107
concessions 26, 112
conflicts 68, 69, 143, 156, 157, 161, 173; agrarian 39; armed 89, 99; intra-community 54; land 55, 137; military 71; social 105–12; social tensions and 33, 56; territorial 52
conglomerates 114
conservation 44, 91–104, 164, 165, 171; importance of 135; policies 166, 173; projects and programmes 89, 146; promoting 167; sustainable 172–3
continuity and sustainability of projects 128–30
Contra rebels 18, 42, 148, 150
contraband 98
co-operatives 133, 141, 145, 148, 172; agroforestry 138, 139, 142–3, 146, 168; organization 149, 150; state 29
copal 71
Corcovado National Park 111, 112, 165
correlation of social forces 140, 141, 146
corruption 65, 69, 99, 137, 139, 140, 144, 166
involving forestry officials 145
state authorities 67, 68
cost–benefit analysis 120, 121
Costa Rica 4–8 *passim*, 12, 17, 46, 49, 83, 89, 115, 123–4, 128–9, 150, 152, 153, 169; agrarian structure/land tenure 35, 36–7, 41, 43; Asociación Indigena 51; Central Valley region 25, 40, 108, 110; coffee production 23; Demographic Association 14, 30; external debt 102; Forest Law (1969) 38, 100; Forestry emergency decree (1987) 114; Forestry Law (1986) 40, 105, 106, 114, 117; fuelwood collection by men 33; General Forestry Directorate 101, 107; ICE (electricity institute) 108–9; IDA (Institute for Agrarian Development) 110, 111; Indians, graziers and colonization 52–4; Indigenous Law 104; Juan Castro Blanco protectorate zone 109; Land and Colonization Institute 53; Law of Land and Colonization 39, 103; Ministry of Agriculture and Livestock 10, 101, 110; National Commission for Indigenous Affairs 52; Peasant Forestry Development programme 108; protected area schemes 90, 105–12, 161, 162; ratio of deforested to reforested land 113–14; recent banana boom 167; sawnwood production 21; tradition of democratic government 171; Tropical Forestry Action Plan 38, 103; Wildlife Division 101; *see also* Arenal; Atlantic coast regions; Borunca; Carara; Corcovado; Hojancha; MIRENEM; Puriscal; San José; Tortuguero
cotton 15, 23, 26, 31, 34, 72, 85, 172
cows 111; *see also* cattle
Cox, S 89, 171
credit 114, 138, 172; cheap, availability of 43; difficult for farmers to obtain 111; facilities 167; in-kind 139; policies 38, 166
crops 23, 29, 74, 110, 119, 123; areas assigned for production 154; cash 23, 51, 64, 73, 110; commercial 39; export 14, 23–6; fallow rotation 20, 43, 52, 56, 72, 73, 84; food 83, 86; forest clearance for 19; non-traditional 64, 125; permanent 85, 86; soil cover too thin to grow 31
CSD (Swiss development agency) 119, 146, 153
Cuba 39
cultures 151; different, clash of 99; disintegration of 158; implications 36; loss of identity 54; perceptions 16, 63,

67; traditional beliefs 63, 67, 159; values 55
Current, D 114, 126, 130, 131, 152
Current, K 53
Cusmapa 85

damage limitation 162
dams 12
Darién region 18, 27, 50, 51; National Park 93
debt 25, 145; debt-for-nature swap mechanism 101, 102, 103; stumpage 157
de Camino, R 152
decision-making processes 16, 44, 62, 150
deforestation: and agro-export development 15–17; and environmental degradation 6–13; and forest protection 158–9; and Indian populations 48–56; and land colonization 82–4; and livelihood 57–81; and shifting peasant agriculture 82–91; and unsustainable development 1–45
demand: for beef 19, 28; for forest products 71; for fuelwood 44; for harvest labour 44; for land and wood 15; for pasture area 20; for railway sleepers 26
democratization 170, 171
demographic trends 4
desert 12
desertification 85
devaluations: *cordoba* 29; *Quetzal* 64, 72
development organizations 146
Diechtl, S 63
diets 72, 85, 155
differential rent 87
discrimination 50, 81
disease: pine borer 66, 69, 136, 158, 159; plant, animal and human 17; tree 66, 154, 155; venereal 49
disempowerment 16
displacement 36, 41, 56
dispossession 34, 36
Dittborn, A 122, 123
diversification 76, 155
division of labour: domestic 62; family 78; gender 77, 81, 121, 122
drainage systems 31, 73
droughts 11, 12, 31
drug addiction 54
dry season 1, 11–12, 29, 42
Dudenhoefer, D 109
Dulin, P 28, 123, 124, 126
dung 72
dwellings 87

ecological and socio-economic characteristics 1–13
economic factors 13; crisis 46, 49, 94, 148, 149; growth 28; stabilization 41, 44–5, 64, 125, 166
ecosystems 117
eco-tourism 101, 147
education 33, 60, 76, 110
EEC (European Economic Community) 110, 145
EIAs (Environmental Impact Assessments) 162
El Mundo 42
El Salvador 9–13 *passim*, 15, 17–19 *passim*, 25–32 *passim*, 48n, 50, 57, 89, 90, 92, 153, 166, 171; agrarian structure/land tenure 35, 41, 43; agroforestry 125–6, 128; CESTA (information booklet) 10; coffee production 23; community nurseries programme 131–3; demographic trends 4; Forestry Law (1973) 96; military activities 167; poverty level 5; reforestation 113, 115, 121; repression 151
elderly people 132
élites 147, 167, 168
Embera Indians 50–1
employment 20, 82, 130; lack of alternative opportunities 164–5; local, important role in sustaining 117; long-term 26; multiple opportunities 81; official statistics 59; off-farm 87; opportunities 29, 71, 74, 81, 121; ratio per hectare of pasture 84; seasonal 55; *see also* wage labourers
empowerment 167, 168, 169, 170
encroachment 52, 53, 55, 152; coffee plantations 151; conflicts over 156; logging companies 144; potential threat of 144
energy 13, 30
environmental degradation 14, 15, 86,

90, 96, 173; deforestation and 6–13; major 22; serious 13, 28, 41; link between population growth and 14, 15, 47
epidemics 49
erosion 9–10, 17, 28, 66, 74, 83, 93; extensive 70; increased 12, 85; levels 13; rain 11, 31, 82; reduced 121; serious 142; soil 12; steep hillsides prone to 73; wind 11, 31, 82; worst instances of 31
ethnic groups *see* Indian populations; indigenous groups
eucalyptus 124, 127
Europe 98, 117, 130
EUROSPECT (mining company) 109
Evangelist sects 67, 156
eviction 152
exploitation 49, 50, 164; *see also* forest exploitation
exports 5, 19, 23–6, 39; agro- 14, 15–17, 18, 24, 84
expropriation 133
'extensive accumulation' 167
extractive activities *see* lumber companies; mining

Faber, D 11, 12, 17, 42, 43
fallow land 82; crop rotation 20, 43, 52, 53, 56, 72, 73, 84; reduced/shortened period 13, 83, 85
FAO (Food and Agriculture Organization of the United Nations) 9–10, 23, 36, 49, 119, 134, 138, 158
farming systems: changes in 83; commercial, improved opportunities for 17; subsistence 50; sustainable 20, 173; swidden 83
farms/farmers: commercial 5, 41; difficult to obtain credit 111; micro 35; more sophisticated 13; peasant 46, 47, 49, 52; small 5, 53, 108; unproductive lands 12
fertilizers 78, 82; natural/organic 64, 72, 73, 135
FETRIXY (indigenous association) 156–8, 170
Figueres Ferrer, José 101
financial agencies 81
Finegan, B 10
fines 65
fishing 52, 54
floods 9, 10, 11, 12, 31, 142
flora and fauna sacrifice 9
Foley, G 28
food 123, 133, 136; areas suitable for production 17; chronic shortages 31; household requirements 73; increased demand for 15; *see also* crops; subsistence provisioning
foreign aid 38, 166; *see also* aid agencies
forest clearance 19, 73, 117
forest cover 19, 40, 60
forest exploitation 49, 55, 71, 81, 137–40 *passim*, 153, 156, 170; attitudes and practices governing 61; benefits of 169; illegal 68, 70, 81; increasing 60; law designed to control 38; sustainable 90, 141, 157; *see also* logging activities
forest fires 54, 70, 135; illegal 104, 136, 152; marked decrease in numbers 157; measures to reduce 138, 155; pine forest areas 10, 142
forest management 123, 141, 145–6; sustainable 14, 54, 62, 116–18, 136, 149, 156, 163
forest products 60, 109; dependence on the sale of 79; illegal exploitation 64; increasing demand for 71; local consumption 71; marketing 123; non-wood 117; use of 149
forest protection 42, 61, 66, 89–173; deforestation and 158–9; important role in the system 63; legislation 37; mechanisms 66; projects and programmes 91; schemes 36, 69, 89; social and political dimensions 160–73; traditional 60, 64
forests: broadleaf 21; colonization of 27, 85; communal 61–9, 78, 117, 136; conversion to pasture 1, 14; dense 10, 57; destruction of 13; family 117; mangrove 1, 10, 11, 32, 33, 95; mixed 57; municipal 69–71; private 69–71, 80; temperate 117; young 155; *see also* agroforestry; deforestation; forest clearance; forest cover; forest exploitation; forest fires; forest management; forest products; forest protection; pine forests; reforestation; SFS
Fournier, L A 52, 89

FPN (National Parks Foundation) 101, 106
fraudulent claims 107
free-fire zones 148
Friedmann, J 169, 170
fruit and vegetables 110
FTN (Franja Transversal del Norte) 18
fuelwood 25, 49, 60, 71, 127, 134, 142; access to 54; crisis 44; gathering 14, 41, 79; household requirements 62; provisioning 72, 77–81, 159; quick-growing species of trees for 119; sale of 149; stoves 77, 136; and urbanization 28–33
Fundación Defensores de la Naturaleza 99
Fundación Guatemala 76
furniture 57, 59, 60, 76

gasoline prices 80
GDP 5, 28
'get rich quick' mentality 99
Gewald, N 30, 31
Ghana 162
GNP (gross national product) 5
goats 110
Godoy, J C 4, 11, 92n
gold miners 107, 111, 112
Golfo Dulce Forest Reserve 111
Goodman, D 34
goods 15, 17, 98
governments 54; cuts in expenditure 49; granted large concessions 25; legislation 37–40; municipal 97; policies 14, 37–40, 46; social programmes 49; *see also* regulations
grain 25, 82, 87, 97, 99, 150; peasant producers 5; sufficient 84
Granados Loarca, J N 60
grasses 74, 83
grassroots initiatives 91, 146, 150–8, 161, 169–73
grave robbing 54
graziers 52–4, 86, 87
Green, G C 49, 92–3, 130
green belt areas 30, 31
'greens' 99
groundwater 12, 22
Guanacaste national park 106, 108
Guatemala 1, 4–6 *passim*, 11–13, 19, 25, 30, 32, 33, 44, 48, 50, 89, 90, 93, 161; agrarian structure/land tenure 35, 41; Altiplano 46, 57, 59, 60, 64, 71, 77; BANSEFOR (national seed bank) 134; coffee production 23; CONAP (conservation agency) 98, 99; DIGEBOS (state forestry corporation) 63, 66, 68, 69, 77, 98, 114, 134, 137; Forestry Law (1974) 96; FUNDAP (NGO) 76; INAFOR (state forestry corporation) 63, 69, 96, 137; National Council of Protected Areas 96; Northern Transversal Strip 39, 40; Protected Areas Law (1989) 96; reforestation and agroforestry projects 133–7; western highlands 28, 57–81; *see also* Huehuetenango; Kekchi; Nahualá; Petén; Quiché; Totonicapán
Guatemala City 77, 98
Guatemala Foundation 75
Guatuso reserve 51
Guaymí Indians 49
guerrillas 18, 41, 99
gum 98
Guzapa Volcano 41

habitats 100, 117
Hall, B 11, 12, 17, 42, 43
'hamburger connection' 19
health 16, 49, 87, 121, 155
Healy, R 37
heart of palm (*palmito*) 129
heat loss 30
Heckadon Moreno, S 19, 23, 26, 29, 33, 38, 82–3, 84, 122, 126, 131–2, 172
Hedström, I 10, 113
Hernández, A 18, 21, 23, 113
hillsides 73, 154
Hojancha 121, 125, 152
Holmberg, J 161
home work system 59
Honduras 1, 4, 10, 11, 17–25 *passim*, 29, 30, 32, 33, 49–50, 89–93 *passim*, 113, 114, 116, 118, 121, 126–8 *passim*, 161, 171; agrarian structure/land tenure 34–5, 37, 38, 39, 41, 43; case study of indigenous groups 170; cattle industry 84; Christian Democratic Movement 140; Coffee Institute 124; COHEP 140; government 44–5; INA (National Agrarian Institute) 95–6; Indigenous Institute 144, 146; Law for

the Protection and Development of Ethnic Groups 152; National Association of Industrialists 140; National Council of Rural Workers 153; National Federation of Honduran Farmers and Ranchers 140; poverty level 5; PROCOINY (development project) 157; social forestry system 137–46; *see also* AMADHO; AMI; Atlantic coast regions; COHDEFOR; FETRIXY; Olancho; Xicaque; Yoro
Houseal, B 50, 51, 52, 53, 151, 170
households: expenditure 72; food requirements 73; indigenous 48; single-parent 150
housing 155
Howard, P 17, 20, 21, 38, 84
Huehuetenango 41, 64, 65, 70, 158–9
human rights abuses 68
humus 11
hunting and gathering 49, 52, 53, 54
Hurricane Joan 10
hydroelectric systems 11, 12, 51, 107, 108

identity 54
ignorance 122
IICA/FLACSO data (1991) 5, 48n
IIED study 116
IIESO report (1976) 73
illegal activities 64, 68, 70, 81, 164; *see also* bark stripping; land; logging; lumber companies; squatters; timber; tree felling
illiteracy 87
incentives 108, 126–7, 134; credit 114; effective 117; financial 101, 117, 152; fiscal 114, 167; food-for-work 132, 135–6; improved, operations of logging companies 173
income 16, 73, 78, 87, 111, 123; additional 129; cash 32, 83, 86; decline in opportunities 74; extremes of 5; inequitable structures of distribution 139; low 168; sources of 29, 79, 86, 98; supplementary 60, 149
indemnification 106, 108
Indian populations 22, 25, 46, 100, 107, 108, 129, 137; acculturation affecting 16; chiefs 145; deforestation and 48–56; discrimination against 81; disintegration of culture 158; indiscriminate shooting of 152; land rights 66; landless 40, 54; pure blooded 158; relations with *ladinos* and state 164; tribes 138; women, marriage with 144; *see also* Chuj; Embera; Guaymí; Kekchi; Kuna; Mayan-Quiche; Mayans; Miskito; Pech; Xicaque
indigenous groups 16, 48, 161, 164, 168; cultures 50, 55, 56; dramatic impact on livelihood 49; failure to respect customary land rights of 165; Honduran case study 170
indio-ladinos 145, 156, 157
industrialists 97
infant mortality 4, 16, 49
inflation 103, 112
infrastructure 54; development of 16, 25, 27–8, 161; economic 16, 97; social 97
INRENARE 10, 93
institutional constraints and biases 122–6
Instituto Agrario Nicaraguense 18
intangibles 121
Inter-American Development Bank 38, 39
Inter-American highway 27
interest groups 97, 146
international peace parks 89
interventions 65, 161–7, 173
investment 38, 51
IRENA (Nicaraguan Natural Resources Institute) 31, 42, 43, 44, 94, 115
irrigation 134
ITTO (International Tropical Timber Organization) 116

Japan 98
Jenkins, J 49, 55
Jones, J 1, 18, 28, 39, 123

Kapp, G 36
Kauck, D 106, 107
Kekchi Indians 99, 151–2
knowledge 16, 162; technical 128
Komer, D 19, 21, 23, 37
Kuna Indians 48, 49, 151, 170, 171, 172

La República 101
labour relations 36, 49
ladinization 53, 158, 159

ladinos 146, 152; colonization by 144; greens versus cowboys 99; incorporated 144; institutions 137; peasant farmers 46, 47, 52; resident 144, 156; *see also indio-ladinos*
land: claims 19; communal 25, 51; conflicts over 137, 155; exploitation of 71; highly inequitable system of distribution 139; illegal transactions 108; improved access to 43; improvements on 87, 104; increased demand for 15; invasions 39; prices 36, 53, 107; private 80; privatization of 165; redistribution 96, 141; speculators 20, 23 taxes on 39; titles 36, 105; unproductive 12, 142; *see also* fallow land; land clearance; land colonization; land concentration; land rights; land-use practices; pasture; tenure
land clearance 29, 36, 44, 166; accelerated processes of 43; for cattle grazing 53; for cotton 32; for pasture 28, 38
land colonization 28, 52, 82–4, 144; agrarian frontier regions 17–19, 25, 34, 36, 38, 98, 123, 164; forest areas 27, 85, 151; government policies which encourage 14; projects 39–40
land concentration 17, 18, 25, 34, 46, 84; acute 36; controls on 157; extreme 35, 57; high levels 43; stemmed 41
land rights 49, 66, 137; failure to respect 165; legal 23, 39; tribal 158
'land to the tiller' programme 43; law (1984) 133
land-use practices 14, 20, 26, 49, 146
landlessness 34, 40, 54, 133
landowners 115, 135; absentee 70, 106; large 97, 106, 151, 152; local 101; potential threat of encroachment by 144; small- and medium-sized 106, 107
landslides 11, 12
Larson, A 10
latifundia/latifundistas 5, 86, 87, 171
lawlessness 99
Leach, G 28
leaf-litter 72, 73
Leal Ruíz, M A 58
left-wing political groups 140
legal titles 70
lending agencies 168
Leon 31, 32, 33
Leonard, H J 1, 4, 11
leucaena 132
licences 63, 68, 70, 80
life expectancy 4
lifestyles 49, 51, 55, 60, 81
Limón 27
literacy 49, 155
live animals 98
live fences 119
livelihoods 57–81, 82, 86; alternative 143, 162; crisis of traditional systems 72; improved opportunities 148; indigenous groups, dramatic impact on 49; threat to 99
logging activities 18, 19, 21–3, 37, 161; companies 41, 54–6, 117, 168, 173; high costs of 115; illegal 111; permits 40; sustainable 116, 117, 118, 173; transnational enterprises 94; uncontrolled 14
Lohmann, L 164
low prices 143
lowland broadleaved forests 116
lumber companies 37, 49, 63, 64, 108, 139, 140–6 *passim*, 153–5 *passim*, 158–9, 168; colonization facilitated by 21; concessions to 37, 118; exploitation 37–8; failure to undertake reforestation 5; illegal transactions with 145; little change in practices 38; national and foreign 98; stumpage fees from 138

McBryde, F 73
McCreery, D 25
McKay, A 28
madereros 106
mahogany 98
Mahony, R 102, 103
maize 12, 75, 78, 84, 86, 136, 143; direct cultivation 71; local diet essentially of 72; prices 73; tree species suitable for alley-cropping of 119; yields 143
Major, M 121, 125
maladministration 139
malaria 49
malnutrition 85
Managua 29, 31, 33
mangroves 1, 10, 11, 32, 33, 95

Mansur, E 9n, 113
MARENA project 94
marginal and inaccessible areas 53
marginalization 16–17, 44, 51, 164; agroforestry co-operatives 168; ethnic groups 50; indigenous populations 49; peasantry 35
María Tecún Ridge 60
marine life 32
markets 15–16, 54, 55; export 5; internal 5, 28; monopolistic 139; prices 120; tourist 76; world 15, 17
marriages 144, 145
Marroquín, A 48n
Martínez, H 28, 30, 33
Marti, Farabundo 151
Mata, A 4, 15, 30
Mayan-Quiche Indians 63, 170
Mayans 77, 98, 144; beliefs 67; culture 50, 63; religious ceremonies 67, 71
MAYAREMA project 99
Mearns, R 28
measles 49
mechanization 18
medicinal plants 49, 54
mejoras 86
Merino, L 4, 15, 30
Mexico 41, 64, 98
migration 20–1, 25, 26, 36, 64, 108, 140, 143, 150; external 4; fuelling the forces underlying 38; internal 41; lowland to highland 52; out- 53, 71–2, 84; patterns 46; rural to urban 30, 32
military activities 41, 71, 97, 148, 167 government 140
minifundia 5, 53, 54, 57
mining operations 49, 51, 52, 108, 161; interests of companies 165; natural resources 99; opening of new access roads for 28; traditional 118; transnational companies 109
MIRENEM (Ministerio de Recursos Naturales, Energia y Minas de Costa Rica) 10, 20, 38, 98, 100, 101, 103, 106, 112, 114
Miskito Indians 41, 55, 151
mixed race 145, 156
mobility 86, 87
mobilization 151, 146, 169–73
modernization 16, 29, 39, 51, 52, 166
Momostenango 70, 79
monoculture 115
Montecristo National Park 96
moon-lighting 166–7
Mora, J A 123–4, 128–9
Morales, R 92
Morazán 9, 41, 42
mountainous areas 1, 12, 13
multiple-use zones 98
multiplier effect 102
multi-cropping 72, 73
multi-track approach 161
Myers, N 151

Nahualá 57–60 *passim*, 69, 71, 74, 76, 77
National Agrarian Institute 152
national and international agencies 130
national parks 92–104, 161; *see also* Chirripo; Corcovado; Darién; Guanacaste; Montecristo; Tortuguero
nationalization 141
Nations, J 19, 21, 23, 37, 89, 171
natural resources 98, 151, 173; destruction of base by rapid population growth 15; highly inefficient in terms of use 142; local knowledge regarding 162; mining 99; more sustainable forms of management 171; pressures on base 52; property rights 114; restrictions on use 108; spiralling depletion of 55
nature 63, 67, 142
Nature Conservancy 101
Netherlands 101
NGOs (non-governmental organizations) 89–90, 100–4 *passim*, 109, 113, 153, 163; conservation 106, 164, 171; environmental 160; international 94; national 94, 99, 119; role of 167
Nicaragua 10, 11, 17–25 *passim*, 30–3 *passim*, 46, 47, 90, 91, 105, 113–15 *passim*, 119, 122, 126, 127, 130, 153, 164; agrarian structure/land tenure 35, 37, 41; coffee production 23; demographic growth rates 4; forest protection 92, 94–5; Forestry Law (1980) 115; GNP 5; Indians and logging companies 54–6; migratory agriculture 84–8; Sí-a-Paz project 94; soil conservation 171; transnational companies and lumber concessions

118; *see also* Atlantic Coast regions; Bosawas; CIERA; IRENA; Nueva Guinea; Pacific coastal areas; Río San Juan; Sandinistas; Segovias; Somoza
Nicaraguan Long Leaf Pine Lumber Co 54
nitrogen-fixing 119, 121
Noriega, Manuel 119
Northern Zelaya 54, 55
Nueva Guinea 39
Nuñez, R 125
nutrients 11, 82
nutritional status 33

oak (*encino – Quercus*) 80
ocote 71
oil 28, 51
Olancho 13, 142; forest reserve 95
olote (maize cob) 78
open-hearth cooking fires 30
Osa Peninsula 111–12, 117
outreach activities 120
outsiders 61, 62, 64, 65

Pacific coastal areas 1, 26; Costa Rica 108; El Salvador 10; Guatemala 11; Honduras 95; Nicaragua 18, 28, 29, 31–2, 39
pajón 74
Panama 4, 8–12 *passim*, 25, 47, 119; agrarian structure/land tenure 35, 36, 38, 39; Canal Zone 10, 28, 93, 94; deforestation and land colonization 82–4; feasibility study in Bocas del Toro 129; forest protection 89, 90; infrastructural development 27–8; *see also* Barra Colorado; Darién; Kuna
Panca 70
Panquix 69, 77
Paqui 62, 68
parcialidades 61–9, 77
Parrita watershed 12, 13
participation 150, 169, 170, 172; greater 171; incentives and 126–7; women's, in reforestation 125
pasture 23, 29, 40, 74, 86; clearing of land for 28, 38, 129; conversion of forest to 1, 14, 16, 43, 38, 83,; demand for 18, 20; employment ratio per hectare 84; encouragement to convert to forest 114; expansion of 35, 107, 111, 138; forest clearance for 19; graziers setting fire to 54; land entirely devoted to 12; nitrogen-fixing trees in 119
pauperization 86
Peace Corps 110, 119, 135, 146
peasantization 140
Pech Indians 152
permits 40, 80, 105
Peru 140
pests 66
Petén 18, 96, 97–9
Peters, R 10
petroleum-based fuels 29
Pickles, D 118
piece-rate work 74, 75
pine borer disease 66, 69, 136, 158, 159
pine forests 1, 63, 64, 71, 138, 142; acidic nature of soils 143; cones 78; dense 57; regeneration 10, 157; resin 139, 142, 143, 145, 153; *see also* Christmas trees; coloured pine; white pine
pit sawing 22
plantations 49, 89, 115, 171; banana 12; cocoa 119; coffee 119, 151
plants 4, 11
ploughing 85
police 65, 68, 112
political parties 145
Poore, D 116, 154
population growth 46; coastal 55; demand associated with 15; environmental degradation and 47; high rates 30; pressure 13, 41; rapid 14, 15, 17
Portico 106
poverty 16, 54, 90, 99, 161; extreme 5, 49, 139; hedge against 17; mass 5
power 97, 167; abuses of 70
precious hardwoods 98
private enterprise 168
privatization 140, 165
production: agricultural 13, 16, 66, 82; agro-export 84; beef 19; cattle 129; coffee 34; costs of 75–6; cotton 172; crop 86, 110, 123, 154; displacement of 17; diversifying 76; food 17, 82, 86, 154; grain 20, 25, 97, 99, 149, 150; petty commodity 71–81; sawnwood production 21; subsistence 107; timber 115

productivity 83; agricultural 9, 28, 31, 79, 82, 85; ranching 84; soil 13
profit 34, 139
project approach 91, 113–30; implementation 131–4
proletarianization 44, 55, 71, 86
property: common 33, 104; private, European concept of 25; rights 68, 107, 114; titles 19; tribal 157; values 107
prostitution 112
protected area schemes 93–112
provisionality 86, 87
public employees 166
punishments 62
Puriscal region 11, 13, 126

quebracho 64
Quiché 41, 70

railroad construction 26
rainfall 1, 17, 42, 53, 70, 88; and erosion 11, 31, 82; force of tropical storms 12
rainy season 11, 12, 42, 54
Rangan, H 169, 170
ranchers/ranching activities 19, 25, 49, 52, 98–9
recession 81, 143; world 41, 43–4
recolonization process 150
Redclift, M 34
redistributive policies 29, 166
reforestation 14, 42, 113–16, 121, 156, 163; adequate rates of 117; appropriateness of 110; failure to undertake 54; incentives for 89, 101, 108, 152; obstacle to 66; projects/programmes 125, 133–7, 141; schemes 110, 128; women's participation in 125
refugees 4, 18, 41, 94
regeneration 10, 62; natural 66, 70, 73, 136, 157
Regional Conservation Units 104
regulations 37, 103, 141; communal 80–1; enforcement of 104, 110, 117; norms and 157; state 81; traditional mechanisms 81; tree felling 38, 63, 77, 133, 138, 163
Reiche, C 30, 31, 120n, 122
Report of the Independent Review (1990) 164
repression 151, 159, 164
reserves 161; biological 95, 100; biosphere 95, 96, 99; biotope 96; conservationist approach 92–104; indigenous 49, 51, 53, 100, 104; national 14; *see also* Arenal; Borunca; Bosawas; Carara; Golfo Dulce; Guatuso; Río Platano
resettlement 148
resource management 162; breakdown of traditional systems 21, 46–91; decision-making processes associated with 16; sustainable 52, 53, 171
resources: access denied to 67; distribution of 126, 170; extraction companies 54; financial 129; limited 117; management of 173; sustainable use of 16; *see also* forest exploitation; natural resources; resource management
Revista del Campo 31, 113
Richards, P 164
Riding, A 40
rights 68: community or customary 81; Indian 50–2; ownership 115; political 172; possession 104, 106; property 68, 107, 114; tribal 158; usufruct 144, 157; women's 62, 165; *see also* land rights
Rigoberto Cabezas project 18, 39
Río Chagres watershed 93
Río Guayapa project 39
Río Lempa 12
Río Pacuare hydroelectric scheme 108
Río Platano Biosphere Reserve 95
Río San Juan 86, 87, 88, 147–50, 166
rivers 142; banks 155; basins 10–11, 12; beds 86; flooding caused by silting 9; falling levels 12; sedimentation of 28
road construction 151
road washouts 12
road-for-timber agreements 27
Rovinski, Y 171
roza (swidden farming) system 83; *see also* slash and burn
Rubinoff, I 10, 12
Ruíz, S 23

Saboval, C 10
salaries 136
Salas Mandujano, E 11–12, 13, 19, 22, 26, 49, 51, 52, 53, 83, 106, 126

San Carlos 148
San Francisco el Alto 69
San Francisco Libre 31
San José 26, 27, 108
San Juan Ixcoy 158, 159
San Mateo Ixtatan 158–9
San Salvador 32
Sandinistas 18, 29, 55, 147, 172; development policy 51; government 87, 148, 149, 151, 167, 169
Santa Catarina 71
Santa Maria de Chiquimula 79
Santiago de Atitlán 152
saplings 115
Sapon 67, 69
Sargent, C 115
Sarmiento, M 28
sawmills 143
Scandinavia 94
Scott, M 35, 36
scrublands 12, 142
SECPLAN 10, 21, 38, 41, 48n, 50, 95, 139, 141, 152
sedimentation 10, 11, 12, 28, 93
seedlings 115–16, 122, 128, 133, 134, 135
Segovias, The 84, 85
semi-desertification 9
Serra, L 41
settlements 85–6, 110, 111
sewage systems 31
SFS (Social Forestry Scheme) 118–33, 137–8, 141–6, 153, 156, 166, 168; demise 161; failure 139
sharecropping 83, 85
Shaw, P 15
sheep grazing 70, 72, 73, 74
SIAs (Social Impact Assessments) 162
Sierra de las Minas 96, 99
Silliman, J 10, 12, 27, 37, 39
silting 9, 31
silvi-pastoral systems 86, 138, 146
slash and burn system 16, 82, 83, 84, 85, 87, 156
slopes 26, 28, 53
small businesses (micro-empresas) 76
smallholders 36
Smith, C 60, 71
social differentiation 35, 55, 74, 81, 86, 170; greater 84
social forces/local level structures 167–9
social relations 82, 83
social services 146
soil fertility 11, 20, 52, 62, 66, 70; activities which would enhance 73; affected by burning 74; declining 12, 13, 82; diminished 31
soils: andeptic 22; changes in organic content 11; compactation due to movement of cattle 11; cover too thin to grow crops 31; erosion 12, 83; inappropriate 86; leaching of 11, 17, 82; loss of quality 117; micro-organisms in 11, 12; not suitable for agriculture 73; pine forest, highly acidic nature of 143; poorer 12, 17, 53; recuperation 52; regeneration 13, 115; sedimentation deposits 12; top layer 11; valley 13; volcanic 11; well-drained 22; *see also* soil fertility
Solis, P 48n, 54, 55
Sololá 58
Someshwar, S 170
Somoza García, Anastasio 18, 29, 39, 51, 54, 87, 147
sparse undergrowth 12
speculators 107
SPN (National Parks Service) 96, 101, 104, 110, 112
squatters 36, 52–3, 54, 105, 106, 107; illegal 39; legislation 40
stabilization policies 149
Standard Fruit company 25, 155
standard of living 50
Stone Container Corporations 118
structural change 147–50
student movement 140
stumpage 129, 138, 141, 144, 157
sub-market prices 106
subsidies 29
subsistence economies 54
subsistence provisioning 64, 123, 143, 144, 159, 164–5; crisis of 71–81; sustainable forms 52; traditional practices 53
sugar 15, 34, 51
sulphur 109
support groups 167
surplus extraction 52, 81
sustainability 20, 128–30, 172–3
Sweden 101, 130
Synnott, T 1, 116, 138

Taiwanese companies 42
Talamanca land 51
tanneries 64, 67
tax evasion 114
technical assistance 123, 125, 143, 166
technology 85; diffusion programmes 125; rudimentary 110; transfer 124
Tegucigalpa 10
tensions 103, 108, 146, 161; intra-community 133; intra-tribal 144, 145
tenure 53, 67, 70, 117, 165; agrarian structure and 34–7; and demise of communal protection mechanisms; 60–71; insecurity of 43, 123; security of 133, 165, 171
terraces/terracing 72, 73, 135, 136
TFAPs (Tropical Forestry Action Plans) 89, 116–17, 164, 169
theft of relics 98
Tico Times 6, 114
Tikal 98
timber 27, 59, 60; for construction 127; illegal extraction 77; increased production 115; processing 112; sale of 70, 157; valuable trees 22; value of 38; volume extracted 116, 117
Tolupan Indians *see* Xicaque
topsoil 12
Tosi, J 106, 107
Tortuguero region 36; National Park 106
Totonicapán 46, 47, 57–81, 115, 162, 164–5; 'Alaska' 74; reforestation and agroforestry projects 133–7; *see also* Mayan-Quiché Indians
tourism 76, 99, 110
trade union organizations 153
traditional beliefs 63, 67, 159
transnational corporations 26, 114, 118, 140, 147; concessions of 112; logging 94; mining 109
tree felling activities 23, 104, 106, 137, 154; artisanal 22; illegal 69, 138; licenses for 63, 68; permits for 105; regulations governing 38, 63, 77, 133, 138, 163; selective 22, 62
tree species: endangered 70; indigenous 124; multiple purpose 49; quick-growing, for fuelwood 119; rigid formulas regarding 132; suitable for alley-cropping 119; *see also* cedar; eucalyptus; leucaena; mahogany; mangroves; oak; pine
trees: broadleaf 4, 21; coniferous 4; cutting, and selling timber for personal gain 70; disease in 66, 154, 155; fruit 72, 132; monoculture planting 115; nitrogen-fixing 119; nursery 134, 156; outlets for products 128; planting initiatives 89–173; quick-growing species for fuelwood 119; respect accorded to, in Mayan culture 63; shade 126, 132; suitable densities in coffee and cocoa plantations 119; valuable timber 22; *see also* logging; lumber companies; pine borer disease; stumpage; timber; tree felling activities; tree species
tribal chiefs *see* caciques
tributary areas 141, 154
Tschinkel, H 123
Tucker, S 167
typhoid 49

underclass 50
underemployment 64, 71, 79, 84
UNDP (United Nations Development Programme) 5
unemployment 64, 71, 79, 84
UNICEF (United Nations Children's Fund) 32
United Fruit company 25, 26, 108, 111
United States 18, 43, 64, 98, 140–1, 168; Alliance for Progress strategy 39; companies 37, 42, 54, 114, 118; conservation organizations 101; demand for beef 19, 28, 38; Department of Agriculture 50; promotion of conservation and war 167; *see also* CARE; Peace Corps; USAID
University of San Carlos 60, 96, 136, 159
UNRISD (United Nations Research Institute for Social Development) ii, 65, 134, 135, 150, 162, 163
unsustainable development 1–45
UPA Nacional 107
upstream activities 128
urbanization 14, 67; fuelwood and 28–33
USAID 9, 10, 13, 37, 42, 50, 94, 97, 99, 115, 117, 119, 134, 136, 146, 167
usufruct rights 144, 157

Utting, P ii, 29, 166

Vahrson, W-G 13
valleys 1, 13
van Buren, A 28, 31, 33
van Sluys, F 19
Veblen, T 47, 59, 60, 61, 170
vegetables 64, 73, 80, 110
Viejo de Najá, Chan K'in 63
Vieman, S 130
Vilas, C 51
violence 41, 68, 81, 164
Vivian, J 16
Vogt, W 9
volcanic activity 13

Wadsworth, F 10
wage labourers 55, 67, 80, 84, 85, 87, 107, 143
wages 20; basic agricultural 78; decline in, public employees 166; differential rates 122; high 67; very low 143
Wall Street Journal 140
war 18, 41–2, 49, 94, 166; promoting 167
water 12, 22, 74, 142; shortages of 10, 31; sources of 54, 74, 134–5; subterranean 54; *see also* rivers; watersheds
watersheds 13, 31, 86, 95, 155; canal zone 10, 94; major 10, 12; protection programmes 93, 137
wealth 5, 90, 139
weeds 82, 83, 116, 135
Weeks, J 34, 35, 36
wells 74
West Africa 164
wheat 73
white pine (*Pinus ayacahuite*) 57, 64, 66, 77, 137
'whites' 108; peasant farmers 52
widows 150
wildland parks 151
wildlife 4, 110; refuges 95, 100, 105, 108
Williams, R 19, 20, 26
wind erosion 11, 31, 82
Winterbottom, R 164
women 55, 68, 122, 132, 150, 155; elderly 78; employment opportunities for 121; fuelwood gathering 33; important role for 156; Indian, marriage with 144; needs 62; participation in reforestation projects 125; rights 62, 165
wood: construction 71; contraband in 98; increased demand for 15; precious 98; products 127, 132, 158; *see also* fuelwood
Woodward, R 25, 27
World Bank 5, 38, 101, 103
World Food Programme 155
WRI (World Resources Institute) 6
Wunderlich, V 49, 51, 52, 53
WWF (World Wide Fund for Nature) 96, 101

xate 98
Xicaque Indians 55, 143–6, 152, 156–8

yields 73, 87, 120, 143, 149; reduced 85
Yodeco 142
Yojoa watershed 95
Yoro 13, 144, 152; Altos Pinares and 20 de Marzo co-operatives 142, 143; grassroots initiatives 153–8; Indigenous Co-operation Project 145; uncertain future of Indian tribes 158; *see also* CAFREYOL
youths 150
yucca 143

Zanotti, J 32

Index compiled by Frank Pert

For Product Safety Concerns and Information please contact our EU representative GPSR@taylorandfrancis.com
Taylor & Francis Verlag GmbH, Kaufingerstraße 24, 80331 München, Germany

www.ingramcontent.com/pod-product-compliance
Ingram Content Group UK Ltd.
Pitfield, Milton Keynes, MK11 3LW, UK
UKHW020201200926
475702UK00004B/561
* 9 7 8 1 8 4 9 7 1 0 2 8 2 *